RECOMME P9-DOD-091

ENGINES

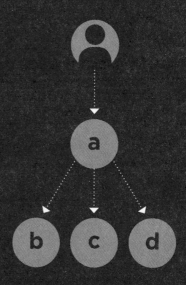

RECOMMENDATION ENGINES

MICHAEL SCHRAGE

The MIT Press | Cambridge, Massachusetts | London, England

This book was set in Chaparral Pro by New Best-set Typesetters Ltd. Printed and bound in the United States of America.

Library of Congress Cataloging-in-Publication Data

Names: Schrage, Michael, author.
Title: Recommendation engines / Michael Schrage.
Description: Cambridge, Massachusetts : The MIT Press, 2020. | Series: The MIT Press essential knowledge series | Includes bibliographical references and index.
Identifiers: LCCN 2019042167 | ISBN 9780262539074 (paperback)
Subjects: LCSH: Recommender systems (Information filtering)
Classification: LCC ZA3084 .S37 2020 | DDC 025.04—dc23
LC record available at https://lccn.loc.gov/2019042167

10 9 8 7 6 5 4 3 2 1

CONTENTS

SERIES FOREWORD

The MIT Press Essential Knowledge series offers accessible, concise, beautifully produced pocket-size books on topics of current interest. Written by leading thinkers, the books in this series deliver expert overviews of subjects that range from the cultural and the historical to the scientific and the technical.

In today's era of instant information gratification, we have ready access to opinions, rationalizations, and superficial descriptions. Much harder to come by is the foundational knowledge that informs a principled understanding of the world. Essential Knowledge books fill that need. Synthesizing specialized subject matter for nonspecialists and engaging critical topics through fundamentals, each of these compact volumes offers readers a point of access to complex ideas.

Who doesn't like—or need—great advice? Who doesn't appreciate the timely and thoughtful suggestion that pleasantly surprises? When I recall pivotal moments in my life—personal and professional—I can't help but remember the impact and influence recommendation and advice had on my decisions. Some suggestions were breathtaking in their awfulness—as if my friend or colleague were totally clueless about me; others were so sharp they startled me to self-awareness and insight I would never have come to on my own. *Why didn't I think of that?*

Of course the "good" in "good advice" is firmly in the mind of the recipient. I've long wondered how my own advice and recommendations were taken, particularly professionally: was I truly being helpful? Or was my counsel more self-indulgent than constructive? How could I have made my suggestion better or more appealing? Pun intended, I've always taken—and given—advice seriously. I know how important—how transformative—the right recommendation at the right time can be.

The right recommendation at the right time is exactly what an Amazon, a Netflix, a Facebook, a Spotify, a Google, a LinkedIn, a Tinder, a TikTok, and a YouTube aspire to. That makes their ambitions transformative. Their massive, high-powered "recommendation engines," also called "recommendation engines," not only algorithmically anticipate what "people like you" desire, they nudge users to explore options and opportunities that might never have crossed their minds.

"We learned what worked and what didn't by seeing how changes in the recommendations helped people find new books," recalls Greg Linden, who pioneered several of Amazon's earliest recommenders and was a great help for this book. "We enjoyed helping people discover books they probably would not have found on their own. It was never about marketing—just matching people to books they would love. But it turns out people do buy more when you help them find what they need."[1]

Indeed. But what happens in a world where that digital device in your hand or the Airpod in your ear instantly and reliably gives you better advice than your best friend? This book describes what recommendations make that world go round.

This is a dispassionate passion project: the purpose and power of recommendation fascinates me intellectually and emotionally. It demands exploration. The way a sharp piece of advice or (seemingly) casual recommendation can change someone's life is awesome and compelling. Reason and feeling converge. Ideally, life and its options get better.

Yet people's relationship with recommendation and advice is as poorly understood as it is underappreciated.

The science—cognitive psychology, neurophysiology, behavioral economics—is not settled. As digital media and algorithmic innovation increasingly define and determine how people "get" and experience advice, we must review and revise what recommendation means.

Bluntly, technology is taking over advice. For some critics, this takeover is hostile; for others, it offers the data-driven gift of intimacy, immediacy, and discovery. Sooner rather than later, most people in most places won't make decisions about what to do or where to go or who to contact without revving up a recommendation engine. They want their recommenders giving them greater confidence and clarity around what they want whenever they want. That's powerful. It charts a future trajectory of human behavior.

More by happenstance than by design, I've been directly and indirectly involved with the rise of recommendation engines since their academic origins at MIT—notably at the Media Lab and Sloan School. While sensing their importance in the moment (my colleagues are smart people), I confess to not thinking big enough. I didn't anticipate clever prototypes hacked with simple statistical rules and semistructured databases scaling so quickly into multidisciplinary juggernauts of big data and machine learning innovation. My imagination failed me. I'd no idea recommendation would evolve from a nifty little start-up feature in software to an organizing principle for commercial—and global—systems design. Stupid me. This book is penance.

Remarkable, isn't it, how trillion-dollar market potential and the power to influence both pop culture and purchasing worldwide can motivate entrepreneurial research? Jeff Bezos, Jack Ma, Reid Hoffman, Katrina Lake, Daniel Ek, Mark Zuckerberg, and Reid Hastings didn't succeed at digital disruption by accident; their organizations delivered recommendations that fundamentally transformed customer expectations and experiences around choice.

This book's purpose is exploring and explaining the "essential knowledge" of this ongoing transformation. History is as important as technology here; humans are as vital as the machines. In fact, recommenders gleefully mash up the most interesting aspects and elements of the mind with the most challenging and controversial components of networked artificial intelligence. They're both "personalization" and "prediction" engines." What, exactly, makes a tune catchy *for you*? Why do you follow *this* suggestion but not *that* one? Recommenders breed inferential interdependence. One can't appreciate the recommendation future without grasping how machines—and their humans—turn data into decisions. There's no escaping either the technology or its underlying mathematics.

Once upon a time, for example, people used maps, a compass, guidebooks, and the stars to navigate the world; today, multimedia medleys of GPS, animated high-resolution displays, and sultry female voices literally suggest people turn left if they want to save seventeen minutes of transit time or veer right should they want incredible views (and, by the way, the restaurant in 3.2 kilometers has the world's best pancakes). Recommendation, advice, and their associated optionality inextricably and increasingly entwine in ways people don't just expect but have come to demand.

Consequently, recommendation offers ways of both understanding the world and understanding oneself. Recommenders prioritize the world's most relevant options and choices for your consideration; those recommendations ostensibly reflect one's tacit and explicit desires: that slice of the world that matters most *to you*. At the same time, you get to see—and decide—whether those options and alternatives resonate with and appeal to who you think (and feel) you are. Do they truly reflect your wants or needs? Do you see—or hear or find—yourself in those choices? Recommendation is both about you "in the moment" and what might come next.

This duality is as provocative and pragmatic as any gedankenexperiment dreamed up by a Descartes or a Heisenberg. Recommendation inspires innovation: that serendipitous suggestion—that surprise—not only changes how you see the world, it transforms how you see—and understand—yourself. Successful recommenders promote discovery of the world and one's self. That's exciting. The shock of self-consciousness is very much a theme of this book. Recommenders aren't just about what we might want to buy; they're about who we might want to become. They're about navigating choice.

As recommendation engines relentlessly colonize global digital ecosystems, those discoveries will go beyond movies, music, books, clothes, videogames, jobs, and people to "friend" and/or "follow" to themes and experiences—*augmented reality? virtual realities?*—we've not yet begun to imagine.

This recommendation-as-innovation ethos subversively infiltrates and influences the questions people behind recommendation engines ask. We have gone from wondering "How can people create more valuable innovation?" to "How can innovation create more valuable people?"

That distinction is subtle but profound. The emphasis shifts from innovation as output to innovation as an investment in human capital and capabilities. That makes recommendation and recommendation engines media, mechanisms, and platforms for creating more valuable people. NB: I'm not saying more productive or more efficient, but more valuable. While this will be discussed at greater length, the importance of recommendation as a way to make people—as consumers, producers, creators, collaborators, colleagues, friends, lovers, or parents—more valuable to themselves and to others can't be overstated.

The book's structure narratively develops and elaborates upon the key themes and issues identified above: Chapter 1 defines, discusses, and details what recommendation engines are and what makes them special. This chapter overviews the social and technical architectures that make recommenders a Web 2.0 network effects platform while identifying the policy and personal risks they pose.

Chapter 2 places recommendation in broad historical contexts ranging from the oracles and astrologers of antiquity to contemporary curators and self-help gurus. Understanding the sociotechnical evolutions of prophecy, prediction, and advice as times and technologies change is key to understanding the role and rule of recommendation engines. This survey chapter describes humanity's quest for great advice.

Chapter 3 reviews the history of recommendation engines themselves; their academic origins, commercial evolution, and multitrillion-dollar global impact. We see the impact of the Netflix Prize on algorithmic innovation and the rapid diffusion of machine learning on recommender systems' technical development. Indeed, even as this book was being written, advances in machine learning radically surged recommender engine power.

Chapter 4 describes how recommendation engines work. It examines the challenge of converting implicit, explicit, and side data into structures that can be algorithmically converted into recommendation. The chapter summarizes key mathematical insights underlying content-based, collaborative filtering and hybrid recommendation engines, as well as addressing how machine learning and deep learning algorithms are transforming the data-rich recommenders future.

Chapter 5 examines recommenders with user experience in mind. Algorithmically excellent recommendations are not enough; people need recommendations presented in clear, accessible, and explainable forms. A picture, for example, can be worth a thousand texts. This chapter also explores the growing influence of behavioral economics, choice architectures, and other research disciplines on recommendation engine design and deployment.

Chapter 6 offers three brief but incisive case studies that illustratively bundle the algorithmic and UX exposition. Sweden's Spotify, China's ByteDance, and America's Stitch Fix—each innovative and successful in its own right—offer superb examples of how recommenders disrupt and redefine business models and markets worldwide.

Chapter 7 concludes with apocalyptic/aspirational visions of possible, probable and inevitable recommendation engines futures. Will nextgen recommenders turn people into meat puppets? Or will they empower new self-awareness and agency that profoundly enhance human potential?

Astute readers will find five interrelated themes informing each chapter of this book's "essential knowledge" trajectory. Individually and collectively, these represent the most important organizing principles for recommendation engines success.

Advice   In the first and final analysis, recommendations are advice; recommendation engines generate advice. Increasingly, that advice is personalized, contextualized, and customized to the individual or group it serves.

Advice can take many forms—text, visual, acoustic, conversational, and augmented/virtual reality—and sophisticated recommenders will present that advice in the format(s) where it is most likely to be valued and used. The future of advice is the future of recommendation engines; the future of recommendation engines is the future of advice. This will be globally true.

Awareness   Depending upon the data that drives them, recommendation engines can create situational awareness of the choices—the options and opportunities—of what to do next. As awareness increases, so does the importance of recommenders for managing it.

Assessment   How well do the recommendations work? Do people follow the advice? Why or why not? Does following the advice reliably lead to desirable outcomes? Objectively and subjectively, do recommendation engines leave people feeling more effective and empowered? Or disappointed and dependent? What metrics are most appropriate for ascertaining success? Amazon and Spotify clearly have different assessment criteria, for example, than health care or professional development recommenders. How does "relevance" get defined and determined? The more intimate and aspirational recommenders become, the more effective those assessment tools need to be.

Accountability   Who bears responsibility for good or manipulative advice? Should good recommenders be liable for bad outcomes? Conversely, should people be sanctioned or punished for ignoring or refusing to follow good advice? Again, the more pervasive, personal and persuasive the recommendation engine, the more accountability matters.

Agency   Agency refers to the power and ability of individuals to act independently and exercise choice. The increasing intelligence and sophistication of recommendation engines, their ability to learn individual and social behaviors, means that they will exert growing influence on individual agency worldwide. Much as identity is shaped by one's family, friends, education, culture, and consumption patterns, recommendation innovation will increasingly shape the agency future.

These five futures can be summarized and synthesized another way: the future of the self. Recommendation engines will continue to transform self-awareness, self-discovery, and self-knowledge. That presents, for better or worse, enormous opportunities and threats. These are discussed in the final chapter. The "essential knowledge" of recommendation engines is better understood through these futures than any particular technology ensemble.

Bluntly, the most difficult challenge in writing this book was not cutting the enormity of this subject down to size but deciding what level of detail would be most relevant and useful. For example, as this was written and reviewed, machine learning evolved from a useful set of algorithmic techniques to a dominant platform for recommendation engine design and production. Yet this can't and shouldn't be yet another book about neural nets and deep learning algorithms any more than a book about cars should be about the important differences between internal combustion engines and batteries.

I want readers to come away not with a technical command of the underlying material but with a sure conceptual grasp of what aspects of technology make recommenders so powerfully effective. There are plenty of resources for deep diving into the technical details; the mission here is for readers to usefully understand how the recommendation engine value chain reliably delivers discovery, novelty, and satisfaction to billions of people worldwide. I want readers to see what fundamentals transcend technologies and markets, and appreciate the "use cases" inspiring innovation. Most of all, I want readers to come away—as I have—in awe of this domain's creativity, diversity, and opportunity. To paraphrase Churchill, we're nowhere near the beginning of the end of recommendation engines innovation. It might fairly be observed, however, that we are at the end of the beginning.

WHAT RECOMMENDERS ARE/ WHY RECOMMENDERS MATTER

As a tool, technology, and digital platform, recommendation engines are far more intriguing and important than their definitions might suggest. Narrow technical definitions shouldn't semantically straitjacket the enormous breadth, range, and impact recommendation engines have on everyday life.

More people around the world are becoming more reliant—even dependent—upon recommendation engines to better advise, inform, and inspire them. Recommenders increasingly influence how individuals spend their time, money, and energy to get more from life. That explains why global organizations ranging from Alibaba to Netflix to Spotify to Amazon to Google invest so heavily in them.

As this chapter illustrates, recommendation engines simultaneously shape and predict their users' futures.

The inevitable Wikipedia defines recommendation engines (and platforms and systems) as "a subclass of information filtering systems that seeks to predict the 'rating' or 'preference' that a user world give to an item."[1]

The *Introduction to Recommender Systems Handbook* declares "Recommender Systems (RSs) are software tools and techniques providing suggestions for items to be of use to a user. . . . In their simplest form, personalized recommendations are offered as ranked lists of items. In performing this ranking, RSs try to predict what the most suitable products or services are, based on the user's preferences and constraints."[2]

A Swedish recommender research paper observes that "A recommender system can be described as a system which automatically selects personally relevant information for users based on their preferences. The problem that a recommender system solves can be defined in many different ways . . . but the most common . . . is as follows: Estimate the degree to which a user will like or dislike items not yet experienced by the user."[3]

Each definition reflects the practical utility and algorithmic underpinnings of what makes recommenders work. They also highlight their essential function: mathematically predicting personal preference.

"At their core," asserts Microsoft researcher Amit Sharma on Quora, "recommendation systems are nothing but 'similarity hunters.' That phrase is deceptively simple

The essential function of recommender systems is mathematically predicting personal preference.

and revealing: The search for similarity—powered by algorithmic innovation—is an astonishingly robust platform for predicting—and proposing—the future.

"Depending on how one defines 'similarity' between two 'items' [or people or groups]," Sharma elaborates, "you can develop a range of [recommendation] applications. . . ."

- Purchase: People who bought X also bought Y

- Experience: People who read/watched/enjoyed X also enjoyed Y

- Location: People who have been at/ate at/stayed at X also went to Y

- Current website: People who come to this website also browse Y

- Education: People who knew about/worked on/ learned/ X also learned Y

- Hiring: People who have skills like your employees

- Recipes: People who cooked X also cooked Y

- Context: People in X mood, at Z time do activity Y more

- Finance: Stocks bought by successful X people

- Popularity: Items popular in the last hour, week, year, event. . . .

- Promotions: People who responded to promotion X should be offered promotion Y

- Social: People/friends are talking about item Y

- Health: People who are healthy do Y more

- Drugs: People with X characteristics respond to drug Y better[4]

Simple similarities—such as "People like you"—can be rooted in whatever aspects, attributes, and attitudes the recommender system recognizes or classifies as relevant. "Similarity" creates predictive pathways to relevance. "Relevance" alternately captures the competition—different kinds of spoons, for example—and the complements— knives, forks, and other cutlery that delineate desirable alternatives. Multidimensional "similarity hunts," however, are merely computational means and mechanisms to a recommendations end. Measurably "better" recommendations for specific users are the ultimate purpose and goal.

Thematically, recommenders aid users along four key dimensions (which, may or may not overlap): they help users *decide* what they could or should do next; they help users *explore* a variety of contextually relevant options;

they help users *compare* those relevant options; and, perhaps most critically, they help users *discover* options and opportunities they might not themselves have imagined. Collectively, this potential help makes recommenders irresistibly appealing to users and developers alike.

This book describes and discusses the computationally matrixed and "machine learned" marriage of data and algorithms that make recommendations so effective. The essential insight is that recommendation engines (r)evolve around a word thus far unmentioned: choice.

In the first and final analysis, recommenders are about choice. The future of recommenders is about the future of choice. The future of choice may well be found in the future of recommenders.

Why Recommendation Engines Matter . . .

Because more people have never had more choices around more opportunities in more domains: because an Amazon Prime Video, for example, offers well over 20,000 movies and videos; because every minute, over five hundred hours of video are uploaded to YouTube; because every day, over fifty million photos upload onto Instagram; because Amazon already carries over three million books while Spotify hosts over twenty million songs.[5] Because great and growing abundance inherently shrink an individual's time and

In the first and final analysis, recommenders are about choice.

attention for thoughtful decision. Because most people realize they need better advice, better suggestions, and better recommendations. Because people who truly want to make better choices increasingly accept that the best recommendations increasingly come from smart machines.

Recommendation engines transform human choice. Much as the steam engine energetically launched an industrial revolution, recommendation engines redefine insight and influence in an algorithmic age. Wherever choice matters, recommenders flourish. Better recommenders invariably mean better choices. Steam powers machines; recommenders empower people. They are the prime movers of their respective eras. They change how work gets done.

That's why Amazon, Alibaba, Google, Netflix, and TikTok are more than mere engines of commerce; they're enablers of individual agency. Their platforms provide instant insight and personalized options to every single user they serve. Their algorithms deliver data-driven suggestions explicitly designed to inspire immediate exploration. Their relentless relevance produces confident curiosity. These recommenders literally—numerically, quantitatively—predict what "people like you"—and *you* in particular—might want or need. That compelling value proposition has infiltrated digital interactions worldwide.

The results speak for themselves: recommender systems influence the video people watch, the books they read, the music they hear, the videogames they play, the investments they make, the friends they meet, the clothes they wear, the food they eat, the restaurants they frequent, the wine they drink, the vacations they take, the news they monitor, the exercises they do, the companions they woo, the products they buy, the cars they drive or hail, the routes they travel, the software they code, the slides they present, the email they send, the classes they take, the art they collect, the babysitters and handymen they hire, the employees they promote, the photos they share, the academic research they review, the gifts they give, the live events they attend, the ads they see, the neighborhoods they live in, the jobs they apply for, the seeds they plant, the pharmaceuticals they take, and—cumulatively—how they actually and practically choose to live their lives.

Wherever mobile devices connect—from Bangalore to Boston to Beijing to Berlin to Bogota—recommenders digitally nudge, advise, and invite more informed decision. Shopping, commerce, and consumption represent only the most obvious examples of their growing influence.

"The recommender system is the key to the success of E-Commerce websites as well as other indexing service providers, such as Alibaba, Ebay, Google, Baidu, You-Tube, etc.," observed data scientists from JD.com, one of China's largest ecommerce sites, in an academic paper describing next-generation recommendation engines.[6] The absence of recommenders virtually guarantees commercial underperformance. Market surveys indicate that customers and prospects worldwide prefer—and act upon—personalized choice. People choose choice. This is true for both emerging and established markets.

One 2019 vendor survey asserts that personalized product recommendation accounted for almost 31 percent of the revenues in the global e-commerce industry. A separate Salesforce study found online shoppers are 4.5 times more likely to add items to their shopping cart and complete a purchase after clicking on any product recommendation. These are not marginal numbers.

Netflix observes that 75 percent of what people watch on the service comes from their personalized product recommendations. (Indeed, recommender data are used to suggest new shows and original programming for the service.) Independent research strongly suggests recommenders directly and indirectly account for roughly a third of Amazon's sales by revenue. Helly Hansen, the Norwegian outdoor apparel store, combined its recommender with weather forecasts, suggesting, for example, rain gear when weather was turning bad in Germany. Hansen's weather-centric advisories soon increased conversion rates 170 percent among regular web shoppers and boosted conversions by over 50 percent for first-time visitors.[7]

Alibaba's self-reported recommender impact was even more impressive. China's largest ecommerce platform disclosed that from Q1 2015 to Q1 2016—a single year—its machine learning–enhanced recommenders more than tripled their impact on the company's gross merchandise volume (GMV) sales.[8] Given that Alibaba's 2016 GMV exceeded half-a-trillion dollars, its recommender systems contribution was enormous.

Perhaps even more than Jeff Bezos—his Amazon counterpart—Alibaba founder Jack Ma made recommendation innovation central to his company's technology and user experience. Recommendation quickly evolved from profitable "add on" to fundamental "organizing principle."

Post-modern, post-industrial digital enterprises depend and build upon recommenders for people and profit. Bezos himself observed in Amazon's early days, "If I have 2 million customers on the Web, I should have 2 million stores on the Web."[9] Recommender systems make "mass personalization" technically and economically possible. They customize even as they scale.

Commercial effectiveness, however, shouldn't obscure a larger individual, societal and global insight: algorithmic recommenders represent a new normal. People all over the world are effectively being trained to expect and respect personalized recommendations. They now assume the data-driven cross-sell, upsell, and serendipitous discovery.

Real-time recommendation is not the occasional opportunistic interruption on one's device but a ubiquitous daily presence. Alternating between useful convenience

Recommender systems make mass personalization technically and economically possible.

and immediate necessity, recommendations inextricably weave throughout the fabric of real-world engagement.

Whether shopping, texting, travelling, working, or socializing, people want their devices suggesting places to eat, photos to share, friends to contact, or shortcuts to take. Tapping, stroking, swiping, or vocally inquiring about a recommendation should instantly proffer appealing options. If not, something's wrong. Crudely put, friends, families, and colleagues may waste one's precious time online, but you can count on data-driven recommenders; they're built for instant relevance.

Behavioral norms and expectations shift along subtler dimensions as well. Truly successful recommenders transcend commercial exchange; they promote personal curiosity and discovery. They're affective as well as effective. Bluntly, successful recommendations make users feel good about their choices.

Selecting that binge-worthy Netflix series or picking emotionally evocative Spotify playlists is not unlike trusting a talented chef to prepare special meals for your family. Of course, a transaction occurs. But what's really happening are little leaps of faith. Users depend on data-enriched algorithmic ensembles to make suggestions that either delightfully affirm or pleasantly surprise.

Skeptics and cynics may disagree but the overarching purpose of recommenders may be less about boosting sales than building trust. As Harmon, O'Donovan, et al.

observed, "The trust that humans place on recommenda-tions is key to the success of recommender systems."[10] Trust is valuable. Trustworthiness is a quality one hopes for—and counts on—from truly knowledgeable friends and experts. Trust invites a risk-adjusted willingness to take a chance. Recommendation invites exploration. Trust builds bonds that create new value chains.

That Amazon book recommendation links to an ob-scure book by the suggested author. A quick scan of an ex-cerpt reveals a footnote to another text. That unexpected book becomes the real find. A recommended YouTube video generates a suggested thumbnail to a scene from a classic Ealing Studio comedy. That celluloid snippet in-spires a Netflix session. That choice opens up a treasure trove of hilariously droll black humor. A Spotify song with a single haunting lyric prompts a Google search to learn about its composer. That link leads to a "hair band" oeuvre that becomes a new Spotify favorite—as well as introduc-tion to a cult TV show that brilliantly sound-tracked the group's music. The junior author of a recommended Re-searchGate paper leads to a SlideShare recommended aca-demic conference presentation where key content—with beautiful graphics—can be quickly clipped and edited (with attribution) into a client sales pitch.

Each offbeat use case turns recommenders into ser-endipity portals. Recommenders become launch pads for self-discovery. As with Google, Bing, and Baidu searches, these journeys can clarify what users are really looking for or starkly reveal what's best left alone.

Either way, the recommender's ultimate impact and influence goes well beyond technical definitions around ranking user preference or utility. "Recommendation cascades" become a new expectation and experience. Changing expectations changes minds.

While serendipity may be happy accident, recommendation-inspired interpersonal engagement is not. A quick screen stroke instantly tweets/Instagrams that Amazon book author with a '*What do you think?*' annotation. Ask friends whether that TV series is worth a three-hour commitment/binge on a rainy Saturday afternoon. That LinkedIn profile can be collegially "slacked" to determine if a credentialed candidate is worth hiring for a crucial crash project.

On a more meta-level, recommenders can recommend which friends and colleagues should receive and review your recommendations. They suggest conversations about suggestions. Not every friend needs sounding out for restaurant or movie suggestions; some colleagues are better at reviewing presentations and papers than others. But sharing recommendations on social media provides a friendly way to further personalize them: *"Yeah, I understand why that movie was suggested but, trust me, what you really want to see is. . . ."*

"Recommendations about recommendations" become another behavioral norm and expectation. Many people enjoy having—and exercising—the option to selectively share specific recommendations: *"These Restaurant/Movie/ Song/News/Travel recommendations should be shared with X"* is fast-becoming part of Sharma's "similarity hunters." Recursively speaking, recommenders help hunt the most relevant similarity hunters.

So recommenders—and "meta-recommendations"— do more than simply select and suggest desirable "items" and "people like you." Time and circumstance help recommenders cultivate more knowledgeable and discriminating consumers. Good recommenders and recommendations virtually educate and train their users. Commercially and culturally, recommenders signal that innovative firms want and value smarter customers.

Their bet: smarter customers make better customers. For long-term businesses, better-informed consumers— trusting consumers—offer, on average, higher "customer lifetime value" than their less discriminating counterparts. Amazon, Booking.com, Facebook, Airbnb, Yelp, and Netflix, for example, know and measurably profit from this empirical insight.

Valued customers make better customers, too. Behavioral economics research and marketing maxims suggest firms want customers demonstrably happy with their choices in movies, clothes, books, travel, and dining. Their customer satisfaction and net promoter scores are higher; their word-of-mouth and social media influence is more persuasive. As a popular pre-Amazon retail television advertisement once proclaimed, "An educated consumer is our best customer." Recommendation engines that don't measurably improve customers—and customer value—inherently underperform those that do.

Clearly, conflicts of interest may prove unavoidable: are recommendations computed with the company's or the customer's best interests at heart/in mind? As will be discussed, managing these competing interests reveals less about technical ingenuity than enterprise values. Empowering users is different than algorithmically exploiting them. That said, making customers better makes better customers.

Facilitating "in the know" consumption, however, is but one side of the microeconomic equation. Recommenders enable value-added production, as well. They make platforms and producers more knowledgeable, selective, and effective. The Facebooks, Alibabas, Amazons, Airbnbs, LinkedIns and Tencents most obviously use recommender data to better segment and serve their customers.

Netflix's pioneering and popular *House of Cards* streaming series relied heavily on recommender data in its design, development, and talent selection. Spotify and Pandora playlist recommenders have similarly proven helpful in identifying new artists for streaming stardom. Amazon, of course, utilizes recommender data to prioritize products to source and bundle. Recommenders are but one part of vast petabyte pools of data these firms use for analytic and predictive power.

But recommenders can also promote greater personal— as well as enterprise—productivity. Marketers and salespeople worldwide use recommenders to plot campaigns and target prospects. Sales Predict, an Israeli start-up founded in 2012 and purchased by eBay four years later, developed analytics focused on recommending high-potential leads and prospects for sales teams. The company quickly discovered that while most salespeople didn't want to be told what to do, they were open to data-driven suggestion.

"Using the word 'recommendations' sounds a little awkward for some business people and we don't exactly want that kind of Amazon-like connection," said Sales Predict cofounder and CEO Yaron Zakia-Or, "but that is, in fact, the direction we went."[11]

Zakia-Or noted that his company learned that typical or average salespeople needed different kinds of recommendations and rationales than top performers. In other words, "top salespeople like you" received different prospect recommendations and supporting material than "ordinary salespeople like you."

Enterprise leaders such as IBM and Salesforce both use and offer such sales recommenders. They're part of the expanding workplace analytics movement so cogently captured by former Google executive and Humu founder

Laszlo Bock's excellent *Work Rules*. Bock's 2015 book described how Google uses data-driven analytics to improve managerial, programmers, and team performance.

But how best to package and present these workplace analytics to boost personal and professional productivity? Spreadsheets, dashboards and dynamic visualizations only go so far. What rhetorical formulation is more persuasive: "You should do this" or "Managers like you consider these next steps. . . ."?

So while many workplace analytic algorithms explicitly offer best or optimal or normative answers, others embrace the recommendation engine (em)power: give executives and employees alike data-enriched, contextually relevant, and algorithmically ranked choice. Don't digitally dictate; advise. Recommenders work quite well for work.

In this crucial regard, recommenders can and should be seen as investments in human capital, that is, as economists assert, investments in "the collective value of the organization's intellectual capital (competencies, knowledge, and skills)."[12] This capital is a constantly renewable source of creativity and innovation capability.

As with education, training, and apprenticeship, recommenders explicitly impart real-world knowledge and advice that makes people more valuable. Technically and economically, recommenders have proven excellent human capital investments for both producers and consumers alike. As recommenders evolve, their contributions to human capital will be even greater.

Jonathan Herlocker, a recommenders research pioneer, enumerated eleven specific ways recommenders empowered their users.[13] This queue of software techniques represents novel technical capabilities for enhancing and augmenting human capital.

Find some good items   A featured list of ranked items found that fit the user's requirements.

Find all good items   A list of all the items that satisfy all the criteria the user set from the item database.

Annotation in text   A list of items that are recommended according to the current context and the long-term user preference. A certain TV series on a certain channel can be recommended according to the user's long-term viewing habits.

Recommend a sequence   A list of items being searched whose sequential interdependence may be interesting and useful for the user.

Recommend a bundle   A list of related items that can work together to serve a purpose better for the user. Typically when you buy a camera, you may consider buying a memory card, a case, and lenses.

Just browsing   For users who browse without a prominent purpose, the recommender system's task is to help the user browse items within the scope that are interesting to the user at that specific browsing session.

Find credible recommender   Some users are skeptical of the recommendation yielded by the system. The recommender system's task is then to allow the user to test the system's behavior.

Improve the [user] profile   The system can take inputs from the user about his or her likes and dislikes, in terms of general, explicit preference information.

Express self   Some users care little about recommendations, but it is important for them to be able to express their opinions and beliefs about certain item. A comment section is where the system can take such inputs, and the satisfaction it creates can serve as a motivation for purchasing the item related.

Help others   Certain users may be even more motivated to leave a full review or rating of the item as a result of their belief this will benefit the community.

Influencing others   Certain users could be exclusively influential, trying to convince other users buying or nor buying the product. Even malicious users can fall into this category.

Whether individually or "ensembled," these recommendation engine options cultivate greater agency and choice for users. As much as any Khan Academy, Udemy, or General Assembly training program, they reflect a human capital ethos. Ongoing innovation in machine learning and human interface design ensures future value propositions that go well beyond Herlocker's first eleven.

Data-Rich Recommendation; Recommendation-Rich Data

More data make recommendation engines more reliable. Recommenders directly benefit from the exponential and combinatorial explosion of datasets worldwide. Estimates understandably—and wildly—vary, but one IBM study asserts daily global data output exceeds 2.5 quintillion bytes—that's 2,500,000,000,000 bytes. The United States alone produces, on average, 2,657,700 gigabytes of internet data every minute.[14] That's overwhelming.

From 2016 to 2017, one marketing research firm asserts, texts-per-minute numbers tripled to over 15.5 million. Every minute, Spotify adds thirteen new songs; YouTubers watch 4.2 million videos; the Weather Channel receives over eighteen million forecast requests; Instagram posts over 46,700 photos; and Google conducts over 3.6 million searches. Does anyone believe these numbers will be going down?[15]

Even tiny slivers of these dataflows may fuel a recommendation engine. But small fractions of enormous quantities have powerful proven predictive impact on efficacy. With the average internet user generating roughly half-a-gigabyte of data every day (and growing), data volumes and velocities invariably improve recommender performance. Reliably anticipating and delivering real-time relevance becomes tractable.

"We've all become willing lab rats, right," former Google data scientist Kai-Fu Lee told a Columbia University engineering school commencement in New York. "Whenever we click and buy and move our mouses to search results, or take an Uber, or buy takeout, or go on an Expedia trip, we contribute a data point—an incredible amount of data."[16]

Personalization is both product and byproduct of the data trails, signatures and "exhaust" people leave in their expanding digital wake. Even casual, "just killing time" interactions speak volumes.

"So, it's a combination of huge amounts of data—never before available—through the internet, with people willingly labeling the data as to what's right and wrong," said Lee. "Labeling the data can be costly and it's critical. So, this large amount of data is inexpensively gathered and stored—because Google came up with all these ways of storing massive amounts of data. They have the new technologies . . . that allow massive data to be collected, labeled, and stored. Now, connect that with the dots of better algorithms."

That marriage—this merger—of big data and better algorithms makes the real-time recommender revolution not just possible but pervasive. Critics such as Harvard's Shoshana Zuboff characterize these massive data marketplace architectures as "surveillance capitalism," which poses its own brand of ethical and cultural concerns around privacy and agency.

But the data/algorithm synergies Lee describes inherently create virtuous cycles of value: more data makes algorithms more valuable; better algorithms make more data more valuable.

This is the essence of the Web 2.0 ethos identified and championed by Tim O'Reilly, the digital publisher and entrepreneurial thought-leader:

> A true Web 2.0 application is one that gets better the more people use it. Google gets smarter every time someone makes a link on the web. Google gets smarter every time someone makes a search. It gets smarter every time someone clicks on an ad. And it immediately acts on that information to improve the experience for everyone else. . . . It's for this reason that I argue that the real heart of Web 2.0 is harnessing collective intelligence.[17]

O'Reilly's 2004 naming and framing has not only stood the digital test of time but has expanded value creation vocabularies. Recommenders have become paradigmatic Web 2.0 platforms, services, and experiences. The more people use them, the more valuable they become. This is globally true. China's digital entrepreneurs have particularly embraced this ethos. So the more people use Amazon's recommendation engines, the more valuable they become to Amazon and everyone who uses it. The same holds demonstrably true for Facebook, Alibaba, Tencent, Netflix, YouTube, Pinterest, LinkedIn, Match.com, eHarmony, Spotify, Quora, and Github.

Every successful "born digital" Web 2.0 innovator, one could observe, enjoys and exploits a recommendation edge.

This is the power and promise—delivered—by harnessing collective intelligence. Successful recommender systems harness collective intelligence in the service of greater individual agency and better choice. Indeed, the definitional differences between search engines and recommendation engines is their data-driven and socially shared personalization. Personalizing Google, for example, would effectively turn its searches into recommendations. Distinctions between search and recommending may digitally dissolve.

As O'Reilly idealistically observed, "The world of Web 2.0 *can* be one in which we share our knowledge and insights, filter the news for each other, find out obscure facts, and make each other smarter and more responsive. We can instrument the world so it becomes something like a giant, responsive organism."

Greg Linden, who helped launch Amazon's recommenders, echoes this sentiment, "I like the idea we are building on the expertise and information of the vast community of the Web. I like the idea that web applications should automatically learn, adapt, and improve based on needs. . . . Web 2.0 applications learn from the behavior of their users. Web 2.0 applications get better and better the more people use them."[18]

Web 2.0 platforms are hardly panaceas. They can be as flawed as the people who run or rely on them. Recommenders are as prone—and vulnerable—to manipulation, misbehavior, and abuse as any other collective endeavor. They suffer technical concerns and challenges. They permit, even encourage, self-indulgence that their critics consider socially malign.

Internet activist Eli Pariser, for example, identified and bemoaned the "filter bubble" impact of recommenders. In this analysis, superior algorithmic personalization becomes more bug than feature: "A unique universe of information for each of us . . . which fundamentally alters the way we encounter ideas and information," he declares.[19] Filter bubbles limit user exposure to conflicting, contradicting, and/or challenging viewpoints. This, Pariser and his supporters argue, facilitates intellectual isolationism and promotes malign political and social polarization.

That said, these are problems of success. This success creates expectations and obligations that may prove difficult to fulfill—or increasingly controversial—going forward. But ongoing innovation, invention, and ingenuity invested in recommendation engines worldwide can't help but impress.

For every identified problem, global communities of researchers and entrepreneurs compete and collaborate to effectively confront them. Those problems and pathologies afflict technology, policy, and commercial opportunity; researchers in academe and on platform are rising to those challenges.

Trust

As discussed, recommenders enjoy their greatest power, influence, and value when trusted by users. Users confident that recommendations respect their best interest are open to the novel, unexpected, and unproven. They're not afraid to take a chance. Indeed, they'll give unknown and untried a shot. They make themselves vulnerable.

That vulnerability creates real risks for manipulation and exploitation. As Dan Tunkelang, who oversaw recommender research at LinkedIn, observed, "The moment that recommendations have the power to influence decisions, they become a target for spammers, scammers, and other people with less-than-noble motives for influencing our decisions."[20] "Shilling" is the broad label for these kinds of manipulative scams.

In practice, technically manipulating recommendation engines is relatively easy. Biasing recommendations to favor one brand or movie or restaurant or person or song over another is not hard. Simply put, betrayal is but a line of code away. Tricking users for one's own benefit requires little ingenuity. But is cheating worth it?

In a *Harvard Business Review* interview, Amazon's Jeff Bezos flatly rejected the business rationale for distorting trustworthy recommendation.[21] He noted and quoted a vendor unhappy with Amazon's willingness to prominently post critical reviews.

> One wrote to me and said, "You don't understand your business. You make money when you sell things. Why do you allow these negative customer reviews?" And when I read that letter, I thought, we don't make money when we sell things. *We make money when we help customers make purchase decisions*. [emphasis added]

Untrustworthy—dishonest or manipulative—recommendations would undermine Bezos's declared commitment to helping customers make purchase decisions. In the longer term, a trust relationship should be more valuable to Amazon than short-term recommendation manipulation. Credible recommendations are currently worth a lot to Amazon, Alibaba, eHarmony, and other Web 2.0 enterprises. Customer Lifetime Value matters more than a transaction or two. In theory, trust is too valuable an asset to squander or subvert.

How—and how much—customers might punish manipulative recommenders is a technically tantalizing question. Companies such as Facebook have come under attack for manipulating feeds as part of experiments and tests. Advertisers are frequently willing to pay extra for greater prominence in listings. Are such self-serving financial arrangement adequately disclosed by the recommenders? Does greater transparency around gray areas cure temptation or yield to it? Again, these are empirically testable questions of ethics, not technology.

Tunkelgang is, of course, correct: the greater the recommender influence, the greater the temptation to manipulate them. Yet the more manipulative and self-serving recommenders are seen to be, however, the less influence they're likely to have. Do recommendation engine incentives favor integrity or exploitation? This promises to be a perennial concern for users, regulators and litigators alike.

Privacy
Privacy, like trust, similarly creates transcendent tensions. As mentioned earlier, Google's Kai-fu Lee noted

The greater the recommender influence, the greater the temptation to manipulate them.

that people have effectively chosen to become data science lab rats—willing participants in statistical and inferential experiments that may prove far more (intimately) revealing than they currently grasp. The recommendations people follow—and ignore—reveal a great deal about who they are.

By design and default, greater personalization requires more personal data and information. Seemingly unrelated datasets may algorithmically blend to yield surprising insights into personal preference. Demographic and locational data—along with time-of-day and heart-rate—might trigger recommendations to invite one of three friends—or colleagues—for drinks (or coffee.) It should surprise no one that researchers and innovators worldwide declare their desire to build recommender systems "that know more about what you want than you do." That strategic goal and aspiration shapes this book's concluding chapter.

With this innovation trajectory, security and confidentiality become even more important. Just how public do people want their dating, financial, medical, and charitable preferences to be? As with health care, "informed consent" becomes more important as recommenders grow more powerful, pervasive, and predictive.

Sparsity

Sparsity is big data's evil twin. Even in digital environments with huge numbers of users and items, most users evaluate just a few items. A variety of collaborative filtering and other algorithmic approaches are used to create "neighborhoods" of similarity profiles. But when users rate just a handful of items then ascertaining tastes/preferences—and appropriate recommendation neighborhoods—becomes mathematically challenging. Sparsity is a "lack of information" problem.

But if users make aspects and elements of their profiles more accessible to recommenders—say, sharing Facebook and Spotify preferences with dating sites or Amazon—all manner of statistically meaningful inferences may be drawn. Sparsity, per se, doesn't vanish but is algorithmically ameliorated by complementary "side data."

This can play important roles managing the "cold start" problem: that is, how to make meaningful recommendations to new users never seen before. (The classic/canonical approach is asking the users to answer a few questions about themselves, as well as display a few of the most popular items as engagement bait.)

Scalability
As user numbers, items, and options grow, recommendation engines need greater computational horsepower to real-time process data. Determining—with ever-higher resolution and granularity—"people like you" and defining ever-subtler features and attributes of items and experiences for ranking and recommendation are hard problems.

As with search, the challenge goes beyond delivering excellent results; it requires delivering excellent results in milliseconds. Complexity and latency are consequently sworn recommender system enemies. Speed (and clever preprocessing) become essential. Computer architecture innovations—at the device and network levels alike—with ongoing enhancements in machine learning, offer best current practice to scaling solutions.

Taking "virtuous cycles" seriously offers an organizing principle* to escape those limitations. Building recommender trust reduces privacy concerns; improving privacy can ameliorate sparsity issues; enhanced scalability improves the real-time recommendations that reinforce trust; more recommendations reduce sparsity.

With ongoing innovation in machine learning, artificial intelligence, sensors, augmented reality, neural technologies, and other digital media, recommendation's reach becomes more pervasive, powerful, and important. The recommendation future promises to be not just more personal, relevant, and better informed but transformative in ways guaranteed to (persuasively) surprise.

ON THE ORIGINS OF
RECOMMENDATION

This chapter takes a sweeping overview of recommendation's remarkably human origins with its remarkable cast of characters. No shortage of genius, imagination, or eccentricity here. The key historical insight: people all over the world—kings and commoners alike—seek new tools, techniques, and technologies in their personal quests for actionable advice.

To be sure, they're not looking to be told what to do; they're asking what they should do next. They want help; they need guidance. They're desperate for insight and foresight. The history of recommendation is the history of how people pursue and perceive advice. What makes that advice good and desirable? The answers to that question describe this chapter's narrative arc.

Recommendation's historical contexts are essential to understanding the dynamics shaping its many futures.

The history of
recommendation is
the history of how
people pursue and
perceive advice.

This history highlights fundamental truths that haven't changed for millennia. Whether offered by gods, wise men, the stars, sacred texts or a roll of the dice, recommendations don't have to be followed, but they must present plausible pathways and possible choices. That's why they matter.

Throughout history—literary, artistic, scientific, military, commercial, and intimate—the recommendations people choose to embrace (or reject) overwhelmingly determine and define their lives.

The most provocative takeaway, perhaps, is that recommendations are engines of introspection and self-discovery. The questions recommendations implicitly raise are as important as the ones they explicitly answer. This essential truth shaped life in Ancient Greece and Imperial China as much as Renaissance Italy or Victorian Britain. The choices people make—the suggestions they follow, ignore, misunderstand, or modify—reveal the duality of who they are and what they want to become.

History suggests that the most effective recommendations inspire curiosity and self-awareness before obedience and embrace. Yes, recommendation's technical methodologies have radically changed; humanity's perennial struggle for self-knowledge and self-command has not. This chapter, though an oversimplification of the origins of recommendations, charts the ongoing conflict and coevolution of recommendation as a means for self-awareness. In reviewing the past, what sources and secrets of good advice matter most? The similarities are remarkable.

In the beginning, recommendation engines were divine. Kings and commoners both sought guidance from the gods. Ancient astrologers around the world charted heavenly influence. Oracles, seers, and soothsayers of classical antiquity interpreted auguries and omens for their concerned supplicants. Desire for divine counsel transcended time, culture, and geography.

In his *On Divination* (44 BCE), the great Roman statesman, orator, and divination skeptic Marcus Tullius Cicero declared, "I know of no people, whether they be learned and refined or barbaric and ignorant that does not consider that future things are indicated by signs and that it is possible for certain people to recognize those signs and predict what will happen."[1]

"There's a very limited set of things that all humans do," observes University of Pennsylvania classical studies professor Peter Struck. "There's eating, walking, and there's divination."[2]

From Latin's *divinare*—"to foresee, to be inspired by a god"—divination is the practice of supernaturally seeking to foretell future events or discover hidden knowledge. The gods—Asian, Babylonian, Egyptian, Greek, Roman—could reveal the outcome of future battles or the perfidy of purported friends. Divination proffers insight as well as foresight.

People needed and wanted signs from their gods; diviners could (seemingly) capture and communicate divine intent. That was the source of their power and influence. Divination drove decision in the ancient world. (Modern day diviners are called "data scientists" and are no less sought after.)

Cicero described two types of divination: one derived by "art," the other determined by "nature."[3] Divinatory "arts" require observation, knowledge, skill, and some form of training—think astrology, extispicy (examining animal entrails), and tasseomancy (reading tea leaves.). By contrast, Cicero's "nature" describes a supernatural divination delivered via visions, dreams, or trances. Think Pythia, High Priestess of the Temple of Apollo—a.k.a. "the Oracle of Delphi."

IBM researcher and science writer Clifford Pickover analogizes artful divination with "deductive" systems where diviners study and interpret signs according to known rules. Ritualistic methods convert seemingly disjointed, random facets of existence into actionable insight. Natural divination, by contrast, is "inductive"; trances, dream-states, and visions transmit the essential advice.[4] Again, the Oracle of Delphi represents history's best-known example of "mediumistic divination."

Whether gripped in a deity's thrall or through artful observation/manipulation of natural phenomena, diviners mediated the essential messages. Divinity-driven "signal processing" generated personalized recommendations for their clientele.

For many ancient and classical diviners, the futures foretold were neither preordained nor fated. Even in a world ruled by gods, people had choice. Knowledge was power; knowledge empowered. Divination could and would help people make better and more informed choices. This was more than mere fortune-telling; this was a chance—an opportunity—to take one's fate into one's own hands. "What should I do?" was as important as "What will happen next?"

In that respect, the ongoing multimillennial transition from "deity-driven" to "data-driven" recommender systems is as remarkable for what hasn't changed as for what has. As radically and rapidly as recommender techniques and technologies have evolved, people's craving for insight persists and proliferates. Today's most significant personal and cultural dimensions of recommendation trace directly back to their ancient origins. Divination may be anachronistic, but it foreshadows the contemporary use of recommendation engines to make decisions. Both look for, listen to, and feel patterns that simultaneously inform and inspire.

Antiquity's seers and *manteis* prove shockingly relevant to contemporary recommender design. As

Divination may be anachronistic, but it foreshadows the role recommendation engines play in contemporary decisions. Both look for, listen to, and feel patterns that simultaneously inform and inspire.

philosopher Rebecca Goldstein observes in her excellent *Plato in the Googleplex: Why Philosophy Won't Go Away*, "classical thought is nothing if not a careful consideration of what constitutes good advice." As Socrates declared, "The unexamined life is not worth living." Recommendation invites self-examination. Accepting that invitation can be transformational.

Essential dialogue and debate about leading a good life, for example, often revolved around how best to consider divine counsel.

This supernatural perspective shaped everyday decision. In a world where Stoics believed in a cosmic *sympatheia*, a holistic and interconnected—might one even say "networked"?—universe, auguries could be anywhere and everywhere. That meant recommendation could be anywhere and everywhere. "It's likely that in antiquity most people practiced or witnessed some form of divination at least once every few days," notes Ohio State classics scholar Sarah Iles Johnston.[5] This was their normal.

"On a scale whose implications we have yet to fully appreciate," asserts Pennsylvania's Struck, "*ancient Greeks and Romans put to use a wide range of technologies that allowed them to hear messages from their gods* [emphasis added]: techniques ranging from—on the one hand—the traditional rituals of authorities such as temple-based oracles, bird- and entrail-reading, and consultation of the Sibylline oracles; and—on the other—private activities such as dream reading, sortation [drawing or casting lots], and astrology."[6]

Divination was truly multimedia. Big deities anticipated "Big Data."

Divine attention typically avoided omniscience or originality. "With very few exceptions," says Struck, "Greek diviners produced incremental advice on tactical matters in the proximate future. These insights may have grand consequences but the insights themselves are small bore. They provide guidance on whether a particular god is angry; whether this or that time is more advantageous for a military attack, or a business deal or a marriage. . . . They do not offer large judgments about the alignment of the universe."[7]

Why is this important? Because kings and commoners didn't seek divine counsel to transform their lives or see their world in new ways; they went because they needed advice. *Should I or shouldn't I? What do the gods have to say?* Divination was a pragmatic choice for serious people. And yet, divination does more than reveal a god's will. Pursuing divine direction is an act of curiosity and humility. Querying an oracle demonstrates vulnerability, not just desire for knowledge, power, and an edge.

For well over a thousand years, Delphi was the *omphalos*—the bellybutton—the center of the ancient world. No oracle in history exerted more influence or enjoyed a better brand. (When the Pentagon's "Wizards of

Armageddon," the RAND Corporation, developed a "collaborative ranking" prediction methodology in the 1950s, it was called the Delphi method.) Antiquity's oracular era came to an end when the Christian emperor Theodosius abolished pagan observance in 395 CE. Divination had become a sin. Nobody saw it coming.

Throughout classical antiquity, "divintech" was medium and method to better "know thyself." Asia's most-enduring divinatory work, the *I Ching*, known also as *The Book of Changes*, has "served for thousands of years as a philosophical taxonomy of the universe, a guide to an ethical life, a manual for rulers and an oracle of one's personal future and the future of the state," declares sinologist/translator Eliot Weinberger. "It has been by far the most consulted of all books" throughout China and East Asia, "in the belief that it can explain everything."[8]

Some five thousand years ago, China's mythical first emperor Fu Xi saw the essential patterns of the universe revealed on the back of a tortoise. This epiphany divine showed the cosmos could be represented by eight essential trigrams—Heaven, Thunder, Mountain, Fire, Water, Lake, Earth, and Wind—each composed of three stacked solid or broken lines. Each line encodes a binary state: Dark/Light, Strong/Yielding. These trigrams reflect the yin and yang duality driving both constancy and change in the universe. What sympatheia was to the Stoics, yin/yang holism symbolizes for the *I Ching*. Everything's connected.

Archaeological records show that diviners of the Shang dynasty (circa 1600 BCE) would fire up tortoise shells or oxen bones and interpret the cracks the heat produced. Such "plastromancy" and "pyroscapulimancy" was historically common throughout Asia. Archaeologists have excavated hundreds of thousands of "oracle bones" inscribed with recognizable hexagrams. Their detailed origins and precise interpretation remain unknown.

The *I Ching*'s hexagrams were ultimately named and their interpretive texts written sometime in 800 BCE.

The *I Ching* itself consists of sixty-four hexagrams, each with its own specific meaning and divination text. Each hexagram consists of two trigrams (which trigram is on top is interpretively important). Each hexagram combines the elemental trigrams into ideogrammatic images and texts describing a unique situation. Each line of each trigram and hexagram is determined by chance—the flip of a coin; the toss of a yarrow stalk. When the hexagram is finally assembled, look it up in the *I Ching* and read its "judgment."

The *I Ching* is thus history's first hexadecimal algorithmically driven "recommendation retrieval" system. What it lacks in personalized data, it more than compensates for in evocative UX and creative content. Each hexagramic judgment—that is, recommendation—advises its elicitor how to confront their chosen question.

The *I Ching*'s impact on Chinese and Asian culture is difficult to overstate. Confucius himself is said to have converted the book from a divinatory text into a philosophical guide.

Even in the West, the *I Ching*'s intellectual influence rivals Delphi's. Artists ranging from Hermann Hesse to John Cage to Merce Cunningham to Bob Dylan openly acknowledge the *I Ching* as a source of inspiration. So have scientific and entrepreneurial geniuses including mathematician Gottfried Leibniz, Nobel physicist Niels Bohr, and Apple's Steve Jobs.

No divintech innovation platform better illuminates the elusive boundaries between divinity, chance, and data-driven reflection than astrology. Its uneasy coevolution with astronomy blends a romantically aspirational science with complex math to explain human behavior. If the tides follow the moon, why wouldn't men's moods? Astral "science" could forecast societal eclipses, not just lunar and solar ones. Like the *I Ching*, astrology offered innovative opportunities to construct explanatory stories around cosmic chance and choice. Calculating celestial trajectories could prove essential to understanding one's own. Phases of the moon might correlate with phases of one's life.

"Astrology is the interpretation and prognostication of events on earth, and of men's characters and dispositions, from the measurement and plotting of the movements and relative positions of the heavenly bodies, of the stars and planets, including among the latter the sun and the moon," states Jim Tester's definitive *A History of Western*

Astrology. "Since astrology proper depends on the charting of the movements and positions of the planets, it could not arise until after the growth of mathematics.[9]

The mathematics are key: the astrological/astronomical ability to successfully forecast an eclipse, for example, understandably lent computational credibility to casting horoscopes. Whether one believes the underlying assumptions or not, converting celestial calculations into individualized horoscopes is nothing more—and nothing less—than a heavenly recommendation engine. All the essential elements are there. But instead of computing mathematical distances between planets to forecast a future, recommenders determine mathematical distances between types of movies and music to predict a preference.

The casting of horoscopes first occurred in Mesopotamia during the Persian occupation in 450 BCE. Babylonian astrologers—creators not just of the Zodiac but of sophisticated mathematics as well—had primarily made predictions for the royal household. This clientele became less lucrative after the Persian conquest. To make a living, Babylonian astrologers began casting personal horoscopes. The first known cuneiform horoscope was cast in 410 BCE. It foretold a child's future from calculating planetary positions at birth.

By the last centuries BCE, notes Ian Bacon, a keen historian of the astrology/astronomy connection, astrology was less science than craft. Its aesthetic was a conceptual/

technical mash-up of Greek astronomical observations and Greek mathematics.[10] Astrology's emphasis and orientation shifted from heavenly exploration to personal prediction. Genethlialogy, for example, takes celestial snapshots of the relative positions of the Sun, the Moon, planets, and the zodiac at the time of birth or conception. Astrologers used those relationships to forecast personal traits as well as individual futures.

Horary astrology attempts to answer a question based on the disposition of the heavens at the moment and place of the questioning. Catarchic astrology casts charts for future endeavors; for example, when best to begin a marriage or business. Catarchic astrology effectively reverse engineers genethlialogy: begin with the desired outcome and determine the optimal moment of heavenly alignment for action. Essentially, different schools of astrology had their own algorithms, techniques, and results. Each school basically developed its own recommender system. Astrology algorithmically foreshadowed today's context-based recommender systems; that is, how moments in time, location, and desired outcomes shape contemporary recommender systems designs.

By the second century BCE, the increasingly powerful Roman Republic came into close cultural contact with Greece. Astrology captured Rome's cultural imagination. A veritable augury ecosystem shaped and shadowed Roman life.

Astrology remained important in the Roman world until the western empire's collapse in the fifth century CE. Medievalists and science historians still debate its influence on the Middle Ages. Astrology survived Christianity's rise in ways the oracles did not. Oracles may come and go; the heavens will always be there. Just look up.

Sparked by rivalry with the Islamic world, astrology enjoyed a twelfth-century European comeback. Court astrologers casting horoscopes and advising royalty abounded By the fifteenth century, nearly every continental court employed its own *astrologus*. Virtually all the leading Renaissance mathematicians—even Galileo—were practicing astrologers.

Here astrology's Renaissance renaissance gives bastard birth to its ultimate disruption. Gerolamo Cardano (1501–1576), a wildly brilliant Italian polymath, proved a controversial but compelling link between astrology's algorithmic pretentions and a "new math" destined to revolutionize the recommendation future: probability. A passionate lover of cards and dice, Cardano was called "the gambling scholar" for good reason. He compulsively bet big. That sensibility shaped his life.

Physician, philosopher, and mathematician, Cardano uniquely straddles divination's occultist past and today's high-octane recommendation engines. He "embraced and amplified all the superstition of his age, and all its learning," declared Henry Morley, one of this Renaissance

Man's many biographers.[11] Cardano's blend of talents and pathologies make him symbol, substance, and cautionary tale of recommendation innovation. He is arguably the Renaissance's most successful failure. (His name also lives on as a popular crypto-currency).

Cardano was astrology's leading public intellectual and innovator—"the most distinguished astrologer of his time."[12] His popular and provocative 1543 book of "celebrity horoscopes" boosted his own eagerly sought celebrity. Astrology made Cardano a star.

But this was not pure sham. As mathematician Philip Davis argues, the Italian's real astrology—while not actually "science"—was rooted in a rigor that merited serious review. Bluntly, the astrologer was a superb mathematician. "Cardano's work represents astrology at its most advanced, most theoretical, most computational hour set forth as a belief and social practice," Davis writes. "This is astrology to which both mathematics and astronomy owe a debt."[13]

Another noted biographer, Princeton historian Anthony Grafton, cruelly but aptly compares Cardano's sophisticated astrology to contemporary econometrics. "At the most abstract level," Grafton asserts, "astrologers ancient and early modern carried out the tasks that twentieth-century society assigns to the economist. Like the economist, the astrologer tried to bring the chaotic phenomena of everyday life into order by fitting them to sharply defined quantitative models. Like the economist, the astrologer insisted, when teaching or writing for professional peers, that astrology had only a limited ability to predict the futures. Like the economist, the astrologer generally found that the events did not match the prediction; and like the economist, the astrologer normally received as a reward for the confirmation of the powers of his art a better job and higher salary."[14] Caveat emptor.

An obsessive self-documenter, Cardano's autobiography reveals a talented self-promoter who relished notoriety as much as celebrity. He was into giving advice.

Cardano enjoyed fame throughout Europe as a physician, mathematician, and engineer. Prolifically publishing astrological, medical, mathematical, technical, and philosophical texts, he was as much a Renaissance recommendation engine as a Renaissance man. Fascinated by all manner of divination and diagnostics, Cardano invented "metoposcopy"—the art of reading an individual's character by studying the broken/unbroken lines on their forehead (sort of a "brow"ser-based *I Ching*). His astrological arrogance, however, directly led to his self-inflicted math destruction.

While traveling in England, Cardano spent "100 hours" casting the young Edward VI's royal horoscope, predicting a long and healthy life. The unfortunate boy proceeded to die within a year. This crushed Cardano's credibility and reputation. That mattered little. With little humility and even less discretion, Cardano subsequently cast a problematic horoscope of Jesus Christ. Rome's Inquisition was not amused. He was arrested and briefly imprisoned. Cardano's star had fallen. His biographers disagree about whether he committed suicide to ensure that he expired on the day his self-cast horoscope had predicted.

Before he died, however, the gambling scholar had secretly pioneered key concepts of probability. His *Liber de Ludo Alae* (*Book on Games of Chance*), written around 1550 but posthumously published in 1663, presented the first known conceptualizations of probability. Cardano defined frequency ratios, as well as formalizing mathematical expectation and deriving power laws for repetitive events. The book drew upon his extensive—and expensive—experiences at dice and cards. Like his astrology, his *Liber* cleverly blended practice and theory, rhetoric, and science. Unlike his horoscopic calculations, however, his gambling scholarship offered credible and testable predictive power. Alas, the brilliant gambler/astrologer had bet on the wrong future.

Probability's rise transformed prediction/recommendation paradigms. Divination shifted from interpreting gods to calculating odds. Instead of asking, "What do the gods say?" successfully compute "What do the odds say?" Luck—*fortuna*—may be a false goddess but, in all likelihood, she is a measurable expectation. Probability, not just causality, becomes a quantitative prism through which to deconstruct nature's secrets. Inference, correlation, regression, significance, and other statistical tools and techniques accelerated the deity-to-data-driven recommendation transition. Like the microscope, telescope, and other innovative scientific instruments, probability became a powerful tool for letting people see and measure what mattered in new ways.

With Bernoulli, Bayes, Fisher, and other probabilistic innovators, statistics grew more descriptive, predictive, and prescriptive. The pseudo-mathematics of auguries, hexagrams, and signs gave way to computationally tractable and empirically testable hypotheses around preference, choice, and advice. Knowing the odds proved reliably more valuable than knowing the gods. "It is remarkable that a science which began with consideration of games of chance," wrote the prescient Laplace in 1812, "should have become the most important object of human knowledge."[15]

As pervasive computation becomes fast, cheap, and easy, probabilistic practices become ubiquitous. Everyone does it. Sympatheia becomes statistical and welcomes data-rich queries to correlate and curve-fit. Recommenders are a logical, inevitable, and technical offspring of probabilistic revolutions.

Yet these revolutions reflect continuity as much as disruption. Probabilistic quantification—no matter how computationally intensive or data-driven—neither subverts nor overturns human fundamentals. Yes, the data and algorithms are different, but the introspection ethos endures. The "gods vs. odds" framing is as relevant to Amazonian recommenders as Delphic oracles: *How do you interpret them? Does knowing them change your mind or behavior? Can you afford to ignore them? Are you looking for advice or affirmation?*

Not all recommendations began with divine sources. Humans have long sought advice from each other, not just as interpreters or go-betweens to the gods.

The recorded history of mortal recommendation and advice begins with proverbs, parables and fables. The *Instructions of Shuruppak* are perhaps the oldest known surviving recommendations. Inscribed in cuneiform script on clay tablets, the ancient (4,700-year-old) Sumerian text offers a king's proverbial advice to his son. These include, "Do not buy an ass which brays too much" . . . "Do not commit rape upon a man's daughter; the courtyard will learn of it," and, of course, "Do not answer back against your father." Indeed.

Shuruppak marks the start of "wisdom literature," the genre embracing the challenges of the everyday. This includes, says philosopher Richard Mussard, "concern guidance for family life and childrearing, guidance about living with others . . . and guidance for bearing life's adversities, injustices, and uncertainties. These works offer answers to questions about the meaning of life—naturally the wisdom writers largely expressed the fruits of their own observations and experience. . . . The literary forms most pervasive are the proverbial sayings and aphorisms."[16]

Like divination, wisdom literature is culturally ubiquitous and pervasive. Yale's Harold Bloom noted that "all of the world's cultures—Asian, African, Middle Eastern, European/Western—have fostered wisdom writing."[17] To be sure, wisdom literature possesses religious components, but suggestion and admonition—not compulsion—are at its rhetorical core. Narrative gives rise to proverbial wisdom.

Aesop's fables, for example, could just as easily and alliteratively been called Aesop's aphorisms. While prevalent in ancient Near Eastern culture, fables appeared in archaic Greek literature as early as eighth century BCE. Aesop himself appears in the fifth century BCE, when Greek authors ranging from Herodotus to Aristophanes reference his fables. According to Plato, Socrates was an Aesop fan.

Aelius Theon, the ancient rhetorician, defined a fable as "a fictitious story picturing a truth."[18] Told at the right time, that truth becomes sage advice. Not unlike the *I Ching*, Aesop's fables provide a recommendation-rich resource for self-discovery. "The Hare and the Tortoise" and "The Dog in the Manger" are not simply children's tales; they speak to fundamental values. "Slow and steady" really can win the race. As with divintech, appreciating fabulous truths requires introspection. People can see themselves—or aspects of themselves—in those brief narratives. The underlying similarities and appeal of these narratives made them ideal delivery systems for recommendation.

In a legend of no small irony, Aesop is said to have traveled to Delphi—for King Croesus, no less—where he reportedly criticized the oracle's pronouncements as too vague and ambiguous. This proved ill-advised. Delphi's outraged citizenry accused him of sacrilege and tossed him off a cliff. Divination was literally the death of the pithy fabulist.

The triumph of the printing press, of course, changed and challenged everything recommendation. "A new technology tends to take as its content the old technology, so that the new technology tends to flood any given present with archaism" said Marshall McLuhan, the Canadian literary critic and pop communications philosopher. "When print was new, it flooded the Renaissance with medieval materials."[19]

Gutenberg's press rediscovered and revitalized old advice and recommendations. For Renaissance merchants, that meant handwritten *avvisi*—advisories and notices—became printed gazettes circulating reviews on what to buy and sell. For the ambitious, "conduct literature," such as Baldassare Castiglione's dialogic *The Courtier* [1528] supplemented "wisdom literature" with advice on how best to impress one's patrons. (Yes, Cardano knew Castiglione.) New markets and new wealth created new media opportunities for recommenders and advisors.

As Elizabeth Eisenstein describes in *The Printing Press as an Agent of Change*, the technology "created conditions that favored, first, new combinations of old ideas and, then, the creation of entirely new systems of thought."[20] Publishing enabled new genres of recommendation and advice to enjoy greater distribution. Recommendation and advice became information people could now see, not just hear. (The rise of Alexas and Siris might mean a pre-Gutenberg return to oral advice and suggestion . . .)

Printing innovation profoundly changed advisory ecosystems. By the eighteenth century, says Eisenstein, "increasingly the well-informed man of affairs had to spend part of each day in temporary isolation from his fellow men";[21] by the nineteenth, "gossiping churchgoers could often learn about local affairs by scanning columns of newsprint in silence as well."[22]

Technologies ostensibly designed to inform invariably end up as media platforms for recommendation and advice. From 1733 to 1758, *Poor Richard's Almanack*, published by Benjamin Franklin, enjoyed enormous popularity precisely because of—*pace* Aesop—its proverbial advice: "He that speaks ill of the Mare, will buy her" and "Fish and Visitors stink after three days."

Poor Richard's Almanack contained all kinds of data—including astrological advice—but its pithy aphoristic recommendations made the publication's fame and fortune.

Between the *Almanack*, *The Way to Wealth* (1757), and his *Autobiography* (1793), Franklin is widely credited as the American Enlightenment's father of "self-help" literature. This genre's massive popularity ultimately superseded and absorbed its "wisdom" and "conduct" literature predecessors. One could just as easily and accurately describe this oeuvre as "recommendation literature." More and more pamphlets and publications sought to package new advice and repackage more conventional wisdoms. Recommendation became a high-growth, high-impact market offer.

Samuel Smiles's *Self Help* (1859) proved a worthy best-selling Victorian extension of Franklin's innovation and inspiration. Filled with biographical examples of perseverance, character, and ingenuity in the face of great odds, the book evangelized the virtues of "self-made men." (Not incidentally, Smiles, also a historian of technology, was an admirer of his irascible contemporary Charles Babbage, often described as the world's first computer scientist.) *Self Help* was a recommendation handbook that both articulated and championed the values of empire. By 1900, it had sold over a quarter-of-a-million copies and been translated beyond Europe into Arabic, Turkish, and Japanese. Like wisdom literature and conduct literature, self-help recommendation transcended geography and culture.

In twentieth-century America, self-help texts and tomes approached their commercial and cultural apogee. Dale Carnegie's *How to Win Friends and Influence People* (1936), published during the Great Depression, quickly became a cultural touchstone—selling over 250,000 copies in its first hundred days. The book has sold over thirty million copies since. (As a young man, American super-investor Warren Buffett took a Dale Carnegie course and reportedly keeps its diploma on his Omaha office wall.) Napoleon Hill's *Think and Grow Rich* (1937) was another Depression-borne self-help classic. Postwar books with titles like *Awaken the Giant*, *Chicken Soup for the Soul*, *The One Minute Manager*, and *The Seven Habits of Highly Effective People* dominated nonfiction best-seller lists.

The essential self-help publishing insight is that the genre's recommendations and advice instantiate a practical philosophy. In his *A Guide to the Good Life: The Ancient Art of Stoic Joy* (2009), William Irvine asserts practical philosophies have two components: "They tell us what things in life are and aren't worth pursuing, and they tell us how to gain the things that are worth having."[23] Practical philosophy impeccably describes the recommender system aspiration. Anyone seeking actionable advice—choices—about what they could (or should) do next had an accessible and affordable resource at their fingertips.

But practical philosophy had another popular publishing tributary. Before *Poor Richard* was even a gleam in Benjamin Franklin's eye, England's John Dunton pioneered a more social and interactive approach to publishing advice. In 1691, he was a philandering thirty-two-year-old bookseller desperately seeking advice on what to do. But he understandably feared exposure. He also realized his awkward circumstance might have broad appeal. Given, as one social critic observed, his "entrepreneurial as well as adulterous spirit was strong," he launched the *Athenian Gazette: Or Casuistical Mercury, Resolving all the most Nice and Curious Questions Proposed by the Ingenious of Either Sex*.

His editorial stroke of genius: opening the publication up to his readers. Dunton pioneered "open source." The gazette was first to publish advice columns based on reader queries. For example, *I knew a gentlewoman who wept the first night she slept with her husband, whether was it joy, fear, or modesty that caused these tears?* And *Why a horse with a round fundament emits a square excrement?* Hugely popular, this was the social media sensation of its day. A young Jonathan Swift was an ardent admirer.

Dunton's magazine—known also as the "Oracle of the Coffee House" [!]—was the progenitor of England's "Agony Aunts" advice columnists, as well as America's own popular "Dear Abbys" and "Ask Prudences." More contemporaneously, his Athenian gazette was print's Quora 0.1.

Dunston's "society of experts," which he called the Athenian Society, proffered their practical advice and philosophies to readers.

These "experts" and "advisors" reflect the other great historical strand of recommendation innovation: curation. "Curation can be a clumsy, sometimes maligned word, but with its Latin root *curare* (to take care of), it captures this irreplaceable human touch," argues Michael Bhaksar. "We want to be surprised. We want expertise, distinctive aesthetic judgments, clear expenditure of time and effort. We relish the messy reality of another's taste and a trusted personal connection."[24]

Curation brings practiced knowledge and sophistication to choice. Ideas and expression lend themselves to the curatorial touch as much as objects or art. The curator—like the diviner, astrologer, or self-help guru and other practical philosophers—offers recommendations and advice that inform choice. As with divination and astrology, curation can evoke—and provoke—the shock of unexpected introspection. This can be seen in everything from travel guides to restaurant reviews and art collections—each with its own heroes, communities, narratives, and genres of advice.

No imagination is necessary to recognize that as choice, complexity, and abundance increase, the need—the demand—for better recommendation and advice intensifies. As people become richer—as choices become more diverse—richer options and opportunities for recommendation emerge.

The origins and histories of recommender systems are as compelling and important as their divine, Delphic, and heavenly antecedents. All the follies, flaws, and failures described in this brief recommendation review are what made recommendation engines not just possible and desirable, but inevitable.

A BRIEF HISTORY OF
RECOMMENDATION ENGINES

From computationally crude and heuristically humble beginnings, recommendation engines rapidly evolved to define and dominate user experience online. The real-time recommendation and personalization that help make Alibaba, Amazon, Facebook, Netflix, YouTube, and Tencent so compelling have their origins in technical innovation and entrepreneurship pioneered in the 1990s. Less than a generation later, seemingly incremental research initiatives disruptively transformed behavioral norms and expectations worldwide. Recommenders make the digital world go round.

The internet's global growth—indeed, the growth of digital media, platforms, and ecosystems—is as much consequence as cause of recommender systems success. Any significant shift in digital technology—ecommerce, mobile devices, cloud computing, GPS, big data, and machine learning—gets rapidly adopted, adapted, and absorbed into recommendations. More and better data invariably means more and better recommendations. Digitizing, formalizing, normalizing, analyzing and scaling word-of-mouth suggestion and review have been complemented by algorithmic innovations designed to promote discovery and serendipity. The pace and process are relentless; their power and persuasiveness impressive.

Ongoing digital innovation drives recommendation; ubiquitous recommendation sparks and spurs digital innovation. This virtuous cycle describes this revolution's past, present, and foreseeable future. Where meta-tags and "thumbs ups!" once dominated, feature engineering and neural nets increasingly rule. This structural shift defines relevance as much for Chinese and Indian consumers as American and European entrepreneurs.

Recommendation engines have become not just core technologies but essential organizing principles for human experience design. What began as academic research tools to manage information overload swiftly became a paradigm and platform for giving people better and more personalized choice.

First-generation recommenders emerged from Xerox's remarkable Palo Alto Research Center (PARC), the birthplace of breakthroughs ranging from object-oriented programming to the Ethernet to laser printing. Apple's Steve Jobs, for example, was inspired by a PARC visit to develop his company's Macintosh computer line. One of the world's most innovative corporate laboratories, PARC was filled with research scientists and engineers who liked to share ideas and build things. Tapestry, announced in 1992, digitally embodied both those norms.

Overseen by David Goldberg, a PARC researcher with a math doctorate from Princeton, Tapestry is widely acknowledged as the first formal computerized "collaborative filtering system." Its purpose was helping Xerox's networked researchers get more value more efficiently from more information. Living in information-rich environments, PARC researchers anticipated that rising tides of digital documents and increased email use would soon overwhelm them. Typical document distribution approaches established multiple listservs that let users subscribe to whatever lists most interested them. Existing technologies couldn't effectively help people prioritize their increasingly complex information needs. Managing digital document flows, they hypothesized, would be enhanced by collaboratively engaging their colleagues.

Tapestry's users would help one another filter by providing explicit reactions to messages. These reactions included sending, receiving, or replying to a document and assigning relevance to it. They could manually annotate or tag their documents for filtering; "read this" or "very smart" or "John asked me to review." Filtering facilitated collaboration.

Existing technologies couldn't effectively help people prioritize their increasingly complex information needs. Managing digital document flows would be enhanced by collaboratively engaging their colleagues.

Tapestry users could also query the system to display articles that matched explicit criteria—keywords, annotations, and, more intriguingly, specific reactions taken by colleagues seeing the article: "Show me all the articles Mary forwarded to this workgroup." Active, personal, shareable, and searchable filtering made Tapestry a value-added inbox-cum-recommendation resource.

Tapestry's ethos, one reviewer keenly observed, was that "eager readers read all docs immediately; casual readers wait for the eager readers to annotate" and "users are provided with personalized mailing list filters instead of being forced to subscribe" to listservs. Better filtration equaled better recommendation.[1]

The technology enabled and integrated three distinct design sensibilities, another PARC reviewer concluded: "collaborative filtering" based on self-declared user interest; rule-based appraisal that automatically categorized and prioritized filtered messages; and visual highlighting that explicitly called attention to the most interesting/relevant message portions. Tapestry enabled serious preliminary efforts to create a digitally dynamic ontology and taxonomy for recommendation.

But the ongoing tasks of manually writing annotations and specifying filters required nontrivial human effort. In real-world tests, Tapestry was neither easy nor automated enough. The vast majority of documents went untagged. Nevertheless, the nascent offering had taken the critical first step by "incorporating user actions and opinions into a message database and search system." Tapestry's interaction model—giving users discretion to pull recommendations out of its database—is known as "pull-active" collaborative filtering.

Intrigued and inspired, Lotus Development researchers adapted collaborative filters to make their popular commercial Notes groupware more collaborative. Their innovation insight was to push interesting Notes content to others who might be similarly interested. These pointers contained a hypertext link to the document of interest, contextual information (title and date of document, name of source database, name of sender), and optional comments by the sender.

Collaboration begat collaboration. Xerox PARC quickly implemented pointers to identify, share, and promote relevant information. This approach is known as "push-active" collaborative filtering. The overarching theme of individuals and workgroups collaboratively ranking and popularizing their shared interests was catching on.

Collaborative filtering systems are self-limiting, however; they require communities of people who effectively know each other. Pull-active systems need participants to know whose opinions they value; they require users who know who might find particular content interesting. Automation was key to overcoming that constraint. Automated collaborative filtering transcended those limitations by using databases of historical user opinions to automatically match participants with similar opinions.

GroupLens, launched by MIT's Paul Resnick and University of Minnesota colleagues in 1992, cracked the automated collaborative filtering code. A research effort built to support the fast-growing Usenet news reader community, GroupLens used "quick ratings" of one to five to evaluate news articles, statistically combine ratings, and generate recommendations based on user profile and preference similarities. With "tongue in recommender cheek," GroupLens called its initiative the Better Bit Bureau until the Better Business Bureau complained.

Though conceptually rooted in Tapestry, GroupLens architecture and algorithms went well beyond their Xerox forerunner. First, the system was built as a distributed network with a scalable architecture that made linking new clients and servers for data-sharing easy. Second, GroupLens developed and deployed an enhanced query engine that could compare ratings, collect each user's queries, make ratings-based suggestions, allow privacy through anonymity, and make predictions. In other words, GroupLens aspired to be more than just a research prototype.

Scalable architecture enabled scalable suggestion; the system automatically identified similar interests between growing numbers of users without them needing to know each other. Tapestry's dependence on community

Automating and algorithmically aggregating collaborative filters profoundly changed how recommendations could grow.

familiarity and disaggregated ratings, by contrast, inherently stunted its growth potential. Automating and algorithmically aggregating collaborative filters profoundly changed how recommendations could grow.

GroupLens pilot studies quickly revealed that, with growing volumes and varieties of users, ratings disagreements increased. Determining average ratings missed the point. To overcome the problem of calculating relevance predictions among diverse groups, GroupLens ran a "correlation engine" that identified participants with comparable ratings profiles. Using these "similarity subsets" to generate recommendations dramatically outperformed methods based on aggregated user average. Their "deceptively simple idea," as they observed in a 1994 paper, was that people who had agreed in the past were more likely to agree in the future.

These "nearest neighbor" or "neighborhood" similarity-hunting algorithms lifted automated collaborative filters to new levels of measurable effectiveness. Suggestion success literally and figuratively depended upon identifying like-minded users.

"In the simplest form, automated collaborative filtering systems keep track of each item a user has rated, along with how well the user liked the item. The system then figures out which users are good predictors for each other by examining the similarity between user tastes. Finally, the system uses these good predictors to find new items to recommend," stated GroupLens collaborators John Riedl and Joseph Konstan.[2]

Historically, however, note that neither GroupLens nor Tapestry was created with consumers in mind. On the contrary, both initiatives grew out of academic research emphasizing interpersonal and group productivity inside organizations—most notably computer supported cooperative work (CSCW), decision support systems, and information retrieval. Collaborative filters were originally intended to transform workplaces, not marketplaces. No matter. Entrepreneurs sensed these nascent technologies had commercial potential far beyond the enterprise. Recommendation innovation could have popular appeal.

The ongoing exponential rise of the World Wide Web and the introduction of the Mosaic browser in 1993, moreover, made clear cyberspace could massively scale. That market reality stimulated further collaborative filtering innovation. The Minnesota GroupLens group broadened its brief, exploring book and movie recommendations with BookLens and MovieLens research initiatives, respectively. But these efforts were no longer unique; collaborative filters were becoming a thing. Mass customizing nearest neighborhoods became a research, development, and innovation theme.

BellKor, the research arm of the then-Baby Bell telephone companies, tested an email-based Video Recommender system for movies. At MIT's media-savvy Media

Lab, Pattie Maes's Software Agents group quickly discovered that the smartest thing software agents could do was make smart recommendations. In 1994, Media Lab grad student Upendra Shardanand's Ringo introduced an email-based music recommender built around profiling user preferences and tastes.

Ringo mooted the cold-start problem by requiring participants to register online and preliminarily rate over 120(!) randomly selected musical artists on a one to seven scale. Participants could submit ratings and reviews and request push notifications and recommendations about music and musicians they might enjoy. Anticipating Spotify Discover by over two decades, Ringo basically emailed personalized playlists to its community members.

A grad school prototype that quickly attracted thousands of users, Ringo begat HOMR—Helpful Online Music Recommendations. HOMR, based on a Media Lab master's thesis, highlighted how automating 'word of mouth' recommendations influenced exploration. Statistically, reliably, and predictably scalable word-of-mouth suggestions in pop culture domains—from music to movies—had enormous commercial appeal. These were social media long before Facebook was even a gleam in Mark Zuckerberg's Harvard eyes.

Collaborative filters could facilitate ecommerce; ecommerce made recommendation desirable. Venture capitalists were intrigued. Ringo/HOMR led to Agents, Inc., an

MIT Media Lab spinoff, that was subsequently renamed and repositioned in 1996 as Firefly. Academic research was now driving postindustrial digital innovation and commercialization.

Indeed, a 1996 academic conference and 1997 paper in the Association for Computing Machinery (ACM) proceedings by Resnick and Hal Varian—who would become Google's first chief economist—formally recast collaborative filters research as recommender systems.[3] The name stuck. Recommenders proved to be less software agent than advisor.

Firmly aligned to World Wide Web protocols, Firefly featured a web interface, an online CD store, and movie recommendations. Firefly also embraced targeted advertising and licensed its recommender technologies to other companies including Yahoo!, as well as Barnes & Noble. (By way of internet history, the Yahoo! search engine launched in 1994 while rival Alta Vista started up in 1995. Google, search's ultimate winner, didn't seriously "soft launch" until around 1997.)

Firefly, which had also invested in user profile privacy and protection technologies, was acquired by Microsoft in 1998 and soon transmuted into Microsoft Passport, the company's authentication protocol. Firefly's boldly innovative recommender efforts subsequently proved as evanescent as its Lampyridae namesake. Firefly soon flickered out.

By contrast, the GroupLensers parlayed their collaborative filtering platform into NetPerceptions Ltd. in 1996. An early customer was a small but exceptionally ambitious 1995 start-up initially launched as an online bookstore, Amazon.com. GroupLens was Amazon's gateway venture to recommendation. Bezos's bookseller soon introduced personalized customer-based book recommendations and bespoke "If you like this author" features that encouraged customers to click to broaden their literary awareness. From its beginning, Amazon took mass customization seriously.

Amazon's aspirations imposed enormous technical challenges on the Twin Cities academic entrepreneurs. Converting research prototypes into robust production systems proved difficult. Collaborative filtering for Amazon users didn't scale nearly as swiftly and smoothly as originally calculated. Correlation engines calculating user similarity could barely keep up. Scalability problems emerged as the number of users overwhelmed the number of ranked items. The imbalance created data gaps that made recommendations harder to compute. Or, as one computer scientist put it, "High sparsity leads to few common ratings between two users."[4]

NetPerceptions revisited its recommendation assumptions. "The big performance breakthrough came," said former NetPerceptions software scientist Brad Miller in a video interview, "when we asked 'What if we compared items against items' instead of 'users with users'?"[5]

"The cool thing about 'item to item' is that we could do a lot of the computation offline [that is, pre-compute those correlations in advance instead of calculating them in real-time]," Miller recalled. "It's a completely different model. . . . We were almost depressed: Why didn't we think of that before?"

"Item-based" similarity recommendations, Amazon and NetPerceptions discovered, could reliably compute better, faster, and cheaper than "user-based" ones. Instead of combing through people's purchasing profiles, measure correlations between product purchases. This unexpected insight significantly improved Amazon's book recommendation experience. As early-days Amazon employee and recommender guru Greg Linden recalled in his blog, "Amazon is well known for their fun feature 'Customers who bought this also bought.' It is a great way to discover related books."[6] "Internally," Linden noted, "that feature is called 'similarities.' Using the feature repeatedly, hopping from detail page to detail page, is called similarity surfing," Linden wrote. "The first version of similarities was quite popular. But it had a problem: the 'Harry Potter' problem. Oh, yes, Harry Potter. Harry Potter is a runaway bestseller. Kids buy it. Adults buy it. Everyone buys it. So, take a book, any book. If you look at all the customers who bought that book, then look at what other books they bought,

"Item-based" similarity recommendations could reliably compute better, faster, and cheaper than "user-based" ones.

rest assured, most of them have bought Harry Potter. . . . This kind of similarity is not very useful," Linden recalled, "[But] after much experimentation, I discovered a new version of similarities that worked quite well. The similarities were non-obvious, helpful, and useful."

Linden added, "When this new version of 'similarities' hit the website, Jeff Bezos walked into my office and literally bowed before me. On his knees, he chanted, 'I am not worthy, I am not worthy.' I didn't know what to say then, and I don't know what to say now. But that memory will stick with me forever."

Does this vignette convey "essential knowledge" of how Amazon's recommenders work? A little. Does it offer essential insight into its founder's commitment to customization and real-time recommendation in Amazon's future success? Absolutely. From the web's beginnings, the most ambitious entrepreneurial founders prioritized and promoted recommendation innovation. Ratings and reviews were surely important but recommendation—often drawing from ratings and reviews—proved the most powerful source of differentiation. Bezos's passion for personalization would be emulated by Netflix's Reed Hastings, Facebook's Mark Zuckerberg, LinkedIn's Reid Hoffman, Stitch Fix's Katrina Lake, Alibaba's Jack Ma, Airbnb's Brian Chesky, and Spotify's Daniel Ek, among others. They were similarly like-minded.

Amazon informed and influenced customers by relentlessly automating word-of-mouth suggestion. The "automation emphasis" changed everything. Indeed, before Linden's "similarities" success, Amazon employed a talented group of editors to write hundreds and hundreds of book reviews for the fledgling site. After testing which approach sold more books, automated recommendation soon replaced professional review. Correlation proved more personal, predictive, and persuasive than criticism (although outsourcing reviews to customers became a hit). Books, of course, were just the similarities canaries in the recommender mine.

Linden featured prominently in another Amazon recommendation innovation that married sales to user experience: "I loved the idea of making recommendations based on the items in your Amazon shopping cart," he declared. "Add a couple things, see what pops up. Add a couple more, see what changes. The idea of recommending items at checkout is nothing new. Grocery stories put candy and other impulse buys in the checkout lanes. Hardware stores put small tools and gadgets near the register."[7] "But here," he observed, "we had an opportunity to personalize impulse buys. It is as if the rack near the checkout lane peered into your grocery cart and magically rearranged the candy based on what you are buying. Health food in your cart? Let's bubble that organic dark chocolate bar to the top of the impulse buys. Steaks and soda? Get those snack-sized potato chip bags up there right away." So Linden quickly hacked up a prototype. He modified the Amazon.com shopping cart page on a test site to recommend other items people might enjoy adding. He started shopping his demo around the company. While reactions were largely positive, the prototype drew the ire of a senior marketing executive. The main objection: those checkout recommendations might distract people from actually completing their purchase. Linden acknowledged that abandoned shopping carts were an ecommerce issue. That invited resistance. "At this point," he recalls, "I was told I was forbidden to work on this any further. I was told Amazon was not ready to launch this feature. It should have stopped there. Instead, I prepared the feature for an online test. I believed in shopping cart recommendations. I wanted to measure the sales impact."

Blocking such low-cost, high-potential experimentation at Amazon was culturally and organizationally taboo. "If you double the number of experiments you do per year," Jeff Bezos has been reported as saying, "you're going to double your inventiveness."[8] This test's results were compelling. "Not only did it win," Linden exulted, "but the feature won by such a wide margin that not having it 'live' was costing Amazon a noticeable chunk of change. With new urgency, shopping cart recommendations launched."[9]

This second tale highlights another critical element of recommendation innovation: real-world experimentation, not just analytic review, drives recommender systems evolution. Data-driven recommendation cultures are experimentation cultures. While this partly reflects an "academic research" ethos, the larger truth is that digital recommendations may rightly be seen as real-world experiments. Recommendations are essentially hypotheses around relevance ruthlessly tested the moment they're proffered. Do users choose to respond to them or not? Tracking the outcomes—these user reactions—is digitally easy.

This captures the "explore vs. exploit" challenge: how should recommenders profitably balance suggestions that reliably exploit what users are known to like versus those that explore novel and potentially serendipitous opportunities? Can recommendation engines learn when to explore and when to exploit? This tension remains one of the most important in recommendation design and innovation.

Netflix, launched in 1997 as a mail-order movie DVD rental service, used recommendation experimentation and innovation to galvanize its struggling business. Repeat rentals were low and slow; most customers wanted only hits. The numbers didn't work. For example, breakeven on a DVD purchase required at least fifteen to twenty rentals (postage not included). Recognizing that his DVDs-by-mail movie plan couldn't profitably scale, the company's cofounder Reed Hastings reengineered the business model.

First, he switched Netflix to a subscription model; this locked potentially fickle one-time renters into longer-term customer commitments where they could view as many

DVDs as desired. The company cleverly created online queues where subscribers could select what they'd next like to see. Simply return the current DVD to get the next. (Not incidentally, this returns feature effectively eliminated any need for late fees, the customer-hated surcharge that so bedeviled Blockbuster, Netflix's biggest video rental rival. Netflix's "No late fees" promise stole market share from Blockbuster.)

Hastings's second innovation proved even bigger. In early 2000, Netflix introduced Cinematch, its movie recommendation engine. Cinematch became a Netflix star by making a value-added virtue of necessity. On the one hand, the software suggested intriguing and unexpected movies for subscribers to watch; on the other, it shifted and smoothed demand for the DVD hits. In other words, Cinematch helped Netflix better balance DVD supply and demand. Recommendation was essential to Hastings's new business vision success.

Cinematch had subscribers rate movies and then segmented them into similarity clusters. A clever algorithmic twist allowed Cinematch to combine preference profiles so couples could get recommendations they might both enjoy. Customers created and managed their own rental queues. Cinematch learned and improved its recommendations as subscribers rated the movies coming in the mail.

"Somewhere in the vicinity of 40% to 50% of the movies that people rent from the site are in some sense mediated by data coming from the recommendation system," Netflix product development VP Neil Hunt told a UCLA Andersen School researcher in 2003. "The more data we collect about user preferences, the better the recommendations."[10]

Diverse, personalized, and non-hits-driven recommendations meant Netflix could efficiently wring better returns from larger fractions of its inventory to serve customers; 80 percent of rentals come from two thousand titles. Even more revealing, over 40 percent of its rentals drew on movies released more than one year ago. That starkly contrasted with 10 percent for typical video stores of the time. In effect, Cinematch was a triple play: it let Netflix simultaneously shape subscriber demand, improve user experience, and increase inventory turns—all while learning more about its customers. Win/Win/Win. Where Blockbuster was a store, Netflix became a trusted advisor.

But viewers weren't the only show business constituency entranced by quality recommendations. TV and motion picture producers were also captivated. As *Wired* magazine contemporaneously observed, "It's the recommendation engine that fascinates independent film producers and Hollywood studios, which see Netflix as a marketing vehicle for off-center movies that are difficult to promote through mass media. Rather than pushing the masses toward what's new—the way Hollywood does today—Netflix pushes subscribers toward titles they're likely to enjoy. Given a large enough customer base, the

Netflix model could even change the way Hollywood develops movies."[11] Hold that thought.

Even as early recommenders redefined ecommerce expectations and experiences for books, movies, and music, ambitious new digital platforms emerged. TripAdvisor started up in 2000, and LinkedIn launched in 2002, while Facebook and Yelp entered the market in 2004. In 2006, Google spent over $1.6 billion to acquire YouTube. Each of these digital innovators would successfully design and engineer their businesses around recommendation. Exponential growth in user choice—colleagues, travel, dining, friendships—demanded radical rethinking in recommendation investment. Then an unexpected competition injected new interest, drama, and urgency in recommendation innovation worldwide.

The Netflix Prize

The Orteig Prize inspired Charles Lindbergh's epochal 1927 solo flight across the Atlantic; the Kremer Prize for human-powered flight launched the Gossamer Condor across the English Channel; the 1994 Ansari X-prize ignited commercial innovation in reusable space craft, which was spectacularly won in 2004. In that provocative spirit, Hastings announced the Netflix Prize to promote recommender research in October 2006. Hastings sought not just to sharpen his recommender's edge but to forge and fashion a whole new blade.

The company released the largest dataset ever made available to the public—over one hundred million movie ratings of one to five stars for seventeen thousand movies from their nearly half million users. The challenge? Any contestant who could predict consumer ratings 10 percent better than Cinematch would win a $1 million prize. Netflix understood that everything about its business got better when recommendations got better. Competitive crowdsourcing was a clever bet on recommendation innovation.

A public relations and technical promotions masterstroke, Netflix's prize attracted literally thousands of academic, industrial, and entrepreneurial entrants from all over the world. It transformed global conversations and collaborations around big data and analytics for people.

"To understand the nature of the task," observed computational social scientist Scott Page, "imagine a giant spreadsheet with a row for each person and a column for each movie. If each user rated every movie, that spreadsheet would contain over 8.5 billion ratings. The data consisted of a mere 100 million ratings. Though an enormous amount of data, it fills in fewer than 1.2 percent of the cells. If you opened the spreadsheet in Excel, you would see mostly blanks. Computer scientists refer to this as *sparse data* [the same sparsity, you will recall, confronting Amazon's first engines]."[12]

The Prize brief, said Page, required predictive algorithms that would successfully fill in those blanks. Contestants were challenged to create a new generation of "collaborative filters" offering more innovative similarity measures between people and between movies. But the premise that similar people should rank the same movie similarly and that each person would rank similar movies similarly still held.

As Page rightly observed, however, "Characterizing similarity between people or movies involves difficult choices: Is Mel Brooks's spoof *Spaceballs* closer to the *Airplane!* comedies or to *Star Wars*, the movie that *Spaceballs* parodied?" Early on, contestants' similarity measures of movies emphasized attributes such as genre (comedy, drama, action), box office receipts, and external rankings. Some models included the presence of specific actors (was Morgan Freeman or Will Smith in the movie?) or event types, such as gruesome deaths, car chases, or sexual intimacy. Later models added data on the number of days between the movie's release to video and the person's day of rental. Questing and testing for novel features and serendipitous correlations dominated competing design frameworks.

Winning the contest, he summarized, required knowledge of the features of movies that matter most;

awareness of available information on movies, methods for representing properties of movies in languages accessible to computers, good mental models of how people rank movies, the ability to develop algorithms to predict ratings, and expertise at combining diverse models into working ensembles. Netflix Prize winners would need to be both creatively holistic and rigorously reductionist.

The competition stimulated enormous amounts of creative energy and algorithmic insight. At the end of the first year, BellKor, a team from AT&T research labs, led all comers. Its one best model processed fifty variables per movie and improved on Cinematch by 6.58 percent. But BellKor had done lots more; by combining their fifty models into a giant ensemble boosted their Cinematch performance to 8.43 percent.

Grasping the increased predictive powers of innovative ensembling, BellKor employed a *Game of Thrones*–type strategy of forging algorithmic alliances with key competitors. One fruitful alliance made combining disparate models easier; another provided perspectives into viewer behaviors. BellKor's bricolage succeeded. Nearly three years later, after barely fending off a last-minute rival ensemble challenge from (really!) Dinosaur Planet, BellKor's Pragmatic Chaos combination triumphantly crossed the 10 percent threshold and won the million dollars. Game over.

"It's been quite a drama," then-Netflix chief product officer Neil Hunt declared at the prize awards ceremony.

At first, a whole lot of teams got in—and they got 6-percent improvement, 7-percent improvement, 8-percent improvement, and then it started slowing down, and we got into year two. There was this long period where they were barely making progress, and we were thinking, "maybe this will never be won." Then there was a great insight among some of the teams—that if they combined their approaches, they actually got better. It was fairly unintuitive to many people [because you generally take the smartest two people and say "come up with a solution"] . . . when you get this combining of these algorithms in certain ways, it started out this "second frenzy." In combination, the teams could get better and better and better.[13]

The ensembling epiphany proved a matrix mathematics masterstroke: it shattered initial hopes and expectations of "silver bullet" solutions to the recommender problem. In reality, reliable recommendation was a complex phenomenon. Humans and machines alike would need to embrace new ways to learn from each other—and the data—to dramatically improve their collaborative abilities to predict what people really wanted.

The competition's success surprised both data scientists and digital innovators. Publicly sharing datasets to attract innovation communities became a best practice.

Humans and machines alike would need to embrace new ways to learn from each other—and the data—to dramatically improve their collaborative abilities to predict what people really wanted.

Indeed, Kaggle—the popular machine learning competition site acquired by Google in 2017—launched the year after the Netflix Prize was awarded.

In 2007, the Association for Computing Machinery (ACM) launched its RecSys conference, a global summit combining the best of academic and industrial research. Recommendation was becoming less a digital subsystem research branch than an emerging ecosystem in its own right. Just as Lindbergh's solo flight across the Atlantic entrepreneurially inspired commercial aviation, the Netflix Prize disruption supercharged digital ambitions to transform the recommendation engine future

"I think the Netflix Prize was immensely important," declared GroupLens innovator John Konstan. "It not only put recommender systems research on the map, but it also interested a lot of excellent machine learning and data mining researchers in the topic."[14]

The prize learnings proved immensely influential as well. In a 2014 presentation reviewing the five years after the award, former Netflix algorithmic engineer Xavier Amatriain—in collaboration with colleagues—detailed findings that had changed innovation trajectories:

• Implicit feedback—click-throughs, views, swell-times, and other measurable user behaviors—has proven better and more reliable to capture preferences than explicit ratings. In other words, user actions reveal more than the ratings they give. Improving implicit feedback instrumentation is crucial to improving recommender effectiveness.

• "Rating prediction" is ultimately not the best formalization of the recommender problem. Incorporating machine learning algorithms that personalize rankings for users offer better approaches to recommending best items for users. Technically speaking, machine learned rankings—e.g., could generate better recommendations bundles than algorithms predicting user ratings. Think of recommendation portfolio management.

• Balancing recommendation trade-offs between exploration and exploitation to engage user is important. What mix of recommendations should inspire curiosity and further exploration vs. those that are "sure things"? That means diversity and novelty of recommendations can be as important as relevance.

• Recommendation is not just a two-dimensional problem of correlating users and items but a multidimensional opportunity embracing contextual elements such as time of day, day of the week, or physical location. Contextual recommendation has won greater research emphasis and investment.

• Users decide to select items based not only on how good they think they are but also on how those decisions may influence or involve their social networks. Social connections—such as Facebook, LinkedIn, and Gmail—can be an excellent source of data to add to the recommendation system. Social media recommender systems can powerfully influence recommendations on other platforms. Over the span of the Netflix Prize competition, for example, Facebook's user base grew from twelve million users to over 360 million users. Access to relevant slices of that data could reduce a lot of sparsity.

• Transparency and accessibility around UX matter. Smart algorithms reliably selecting the most contextually relevant items for users isn't good enough. Those items need to be presented in formats users can appreciate and usefully engage.

• Relatedly, reasons and rationales for the recommendations must be readily explainable. Explaining the recommendations usually makes it easier for users to make decisions, increasing conversion rates and leading to more satisfaction and trust in the system. Providing explanations leads to better understanding and a "sense of forgiveness" if particular users dislike particular recommendations. How to automatically generate and present system-side explanations has attracted serious research interest.

To hear serious researchers and investors in Silicon Valley or Shanghai tell it, the Netflix Prize's impact on data-driven development in recommender, platform, and machine learning is difficult to overstate. Seemingly oddball machine learning theories got stress-tested in real-world contexts. Once-contrarian and controversial discoveries have become conventional wisdoms. The competition fostered innovative ensembles of recommender innovators and entrepreneurs as well.

The great irony—even perversity—of the Netflix Prize is that the company ultimately didn't use the winning recommender algorithms or ensembles to transform Cinematch. Times and technology had disruptively evolved. The mailbox DVD was dying; innovation in on-demand higher-bandwidth cloud computing encouraged Netflix to embrace web-streamed video distribution. Subscribers loved it, but the delivery technologies, UX, and immediacy were all profoundly different. The Cinematch criteria and queue management assumptions that had made Netflix's recommendations-oriented business model so successful devolved into anachronism. *Sic transit gloria commendaticiis*.

"When we were a DVD-by-mail company and people gave us a rating, they were expressing a thought process," said Netflix's Amatriain. "You added something to your queue because you wanted to watch it a few days later; there was a cost in your decision and a delayed reward. With instant streaming, you start playing something, [if] you don't like it, you just switch. Users don't really perceive the benefit of giving explicit feedback, so they invest less effort."[15]

His Netflix colleague, Carlos Gomez-Uribe, agreed: "Testing has shown that the predicted ratings aren't actually super-useful, while what you're actually playing is. We're going from focusing exclusively on ratings and rating predictions to depending on a more complex ecosystem of algorithms."[16]

A "complex ecosystem of algorithms" sounds like "ensemble." What kind of data do algorithmic ecosystems/ensembles process to improve recommendation quality? Netflix online now monitors when subscribers pause, rewind, or fast forward; what days subscribers watch content (the company has found people typically watch TV shows during the week and save movies for the weekend); the date subscribers watch; the time; the location (by zip code); the device; and, of course, the ratings. Monitoring those behaviors in a DVD world was impossible.

The ability to stream video content to people's devices elevated the implicit feedback imperative. Although the Netflix Prize failed to refine or revolutionize yesteryear's Cinematch, it presciently positioned Netflix to profit from its new on-demand digital delivery platform. The competition gave Netflix a clear and compelling case for rebuilding its recommendation future.

Even better, this new wealth of data-driven customer insight—observing, for example, which show recommendations inspired binge-watching—empowered Netflix not just to recommend content but to create it. Recommenders can, indeed, be double-edged digital swords. The same data and algorithms Netflix used to help subscribers decide what shows to watch could be repurposed to help producers decide what shows to produce. Recommendation engines could be as valuable for creative development as for effective distribution.

In 2011, Netflix outbid HBO and AMC for the right to produce a US version of the British TV hit *House of Cards*. Recommender-derived analytics were central to the company's willingness to make a $100 million bet to get into the business of original production. Netflix's recommenders had become risk management tools. According to the Kissmetrics blog, Netflix knew many of its subscribers watched the David Fincher–directed movie *The Social Network*—about Facebook, no less—from beginning to end. Netflix knew the British *House of Cards* enjoyed considerable support. But in addition Netflix knew fans of the British *House of Cards* also watched Kevin Spacey films and/or films directed by Fincher.

In essence, recommenders could advise Netflix programmers about what shows would most likely win subscriber curiosity, loyalty, and "binge budgets." "Because we have a direct relationship with consumers," said one Netflix executive, "we know what people like to watch and that helps us understand how big the interest is going to be for a given show. It gave us some confidence that we could find an audience for a show like *House of Cards*."[17]

Until its star's sexual assault scandals, *House of Cards* proved remarkably successful as both artistic and financial investment. But Netflix's "recommender effect" goes well beyond analytics for original content design; it permeates every content acquisition the company considers. "Netflix seeks the most efficient content," declared a former Netflix product engineering VP. "Efficient here meaning content that will achieve the maximum happiness per dollar spent. There are various complicated metrics used but what they are intended to measure is happiness among Netflix members. How much would it go up if Netflix licenses, say, *Mad Men* vs. *Sons of Anarchy*?"[18]

Without its recommendation infrastructure, those questions would be largely hypothetical. But the company's culture and innovation investments ensured those questions have real, reasonable, and profitable answers. Television and movie producers, networks, and distributors all over the world analyze Netflix's analytics to inform their own decisions. The Netflix model has indeed transformed Hollywood and its economics.

Google's YouTube similarly found recommendation crucial to UX appeal and business success. But its business model and technology required a fundamentally different approach. Supported by advertisers, not subscribers, and featuring uploaded videos that often lacked anything remotely resembling Hollywood production values, the internet's largest video site originally privileged search over recommendation. That quickly changed. Search simply didn't get the job done for either viewers or advertisers.

Borrowing programming metaphors from television, YouTube initially asked viewers to subscribe to channels. The results underwhelmed; average viewing times per user didn't budge.

A 2010 RecSys paper on YouTube's nascent recommender system, which then strongly resembled Amazon's "item to item" approach, outlined the technical issues confronting this anti-Netflix innovator:

Recommending interesting and personally relevant videos to [YouTube] users [is] a unique challenge: Videos as they are uploaded by users often have no or very poor metadata. The video corpus size is roughly on the same order of magnitude as the number of active users. Furthermore, videos on YouTube are mostly short form (under 10 minutes in length). User interactions are thus relatively short and noisy . . . [unlike] Netflix or Amazon where renting a movie or purchasing an item are very clear declarations of intent. In addition, many of the interesting videos on YouTube have a short life cycle going from upload to viral in the order of days requiring constant freshness of recommendation.[19]

The early system relied on a combination of weighted explicit and implicit feedback: "To compute personalized recommendations we combine the related videos association rules with a user's personal activity on the site: This can include both videos that were watched (potentially beyond a certain threshold), as well as videos that were explicitly favorited, "liked," rated, or added to playlists. . . . Recommendations [are the] related videos . . . for each video [the user has watched or liked after they are] ranked by . . . video quality . . . user's unique taste and preferences [and filtered] to further increase diversity."[20]

Bidding to boost viewer engagement, YouTube in 2010/2011 made "video views" its key performance indicator. That is, the company wanted more viewers watching more videos more often. So the technical leadership quietly enlisted Sibyl, a nascent Google machine learning system, to train its recommender algorithm to maximize clicks. Boosting click-through rates was its optimization mantra.

This was, at the time, arguably industry's most significant machine learning application to recommendation.

YouTube's recommendation engine, a quick study, quickly delivered outstanding results; viewer clicks surged. Many more people watched many more videos. The numbers were terrific.

But be careful what you wish for: YouTube discovered its cutting-edge software had trained a manipulative click-bait monster. "We were so focused on the views," said You-Tube's Christos Goodrow, an engineering director. "When we launched, it was great as the views went up a lot and we were very happy about that. . . . But as we started to look at [user] sessions, we noticed that some sessions would have many more views but they actually seemed to be the worst experiences for the viewers."[21] YouTube's recommender had trained itself to be more self-serving than user-centric. That proved pathologically counterproductive.

Instead of suggesting the best or most relevant videos for viewers, the algorithm highlighted the best and most relevant videos for generating viewer clicks; it played tricks to get clicks. "We were looking at a session in which a user would have searched for a boxing match. At the top of the search results, they found what looked like a video of the boxing match. The thumbnail was very compelling—but when you click it what you saw was not the actual video of the fight, but someone talking about the fight. The user would watch the video for a minute or 30 seconds, and realize [it] would not have any footage of the actual fight." The more irritated users clicked through multiple videos, the more those increased views were interpreted as signals of success. The results were decidedly perverse.

"This system was like catching a rocket booster for your product and whatever direction you point it in, it was going to go much faster in that direction," Goodrow acknowledged. "So if you have not pointed it in the right direction, you are going to be way off very quickly and that is where we were headed." Indeed.

YouTube flipped its critical success criterion in early 2012. Rather than count views, the algorithm measured time. Before the change, YouTube would only track view lengths up to thirty seconds; the shift meant monitoring timeframes as long as two or three minutes. "Our goal is we want users to watch more and click less," said Goodrow. "This is better for users because it takes less clicking to get to the video you want to watch." The machine learning system now trained the recommender to recognize and reward videos that won people's time. It worked, but the transition proved excruciating.

Viewership plummeted over 28 percent and took almost a year before climbing again. Sloughing off the video parasites and predators who had successfully gamed YouTube's clickbait engine took months. Eventually, YouTube said, average viewing sessions on mobile shot up more than 50 percent, to over forty minutes. Hours watched on mobile doubled. Watch time on YouTube reportedly grew 50 percent a year for the next three years.

"Our aspiration is to be the Holy Grail of recommendation systems," Goodrow said about the transition. "Somehow we have to figure out a way to help viewers find videos that they didn't even know would be on YouTube. We don't know how to do that very well yet—but that is where our aspiration is—how can we make the platform so that the user does not have to put in any energy and yet somehow we show them some suggestions that they actually love."

YouTube would soon turn to another machine learning platform, Google Brain, to get closer to that Holy Grail. Identifying the right criteria to build recommendation engines around has historically proven to be both a business and technical challenge. Consider Match.com, the pioneering internet dating site that originally launched in 1995. It took over a decade for its leadership to take recommendation seriously. After becoming executive vice president and general manager of Match.com's North American operations in 2008, Mandy Ginsberg quickly set about building an analytics team to overhaul the company's primitive matching algorithms.

Prior to Match, Ginsberg had worked at i2, a supply chain management software company with an excellent quantitative reputation. She poached from her old firm. "I brought over a bunch of people who I thought could help solve one of the most difficult problems out there, which is how to model human attraction," she said.[22]

Her key i2 hire was IIT Mumbai graduate and chemical engineer, Amarnath Thombre, who became Match.com's VP of strategy analytics. "The one thing I knew was numbers and analytics, so we started building a numbers team here," he told the *Financial Times*. "It's just supply and demand. The same principles work, no matter what kind of numerical problem you're solving."

Not quite. Match.com confronted recommender design challenges a Netflix or YouTube did not. Match's recommendations had to explicitly embrace and evaluate reciprocity. "Even if you like *The Godfather*," Thombre observed, *The Godfather* doesn't have to like you back. The whole problem of mutual matching makes the problem 10 times more complicated."

That complication created algorithmic and computational opportunity. Before Thombre's analytics team came on board, "matches were based on the criteria you set. You meet her criteria, and she meets yours, so you're a good match," he said. "But when we researched the data the whole idea of 'dissonance' came into focus. People were doing something very different from the things they said they wanted on their profile. . . . We are so focused on behavior rather than stated preferences because we find people break from their stated preferences so often."

Match.com began "weighting" and calibrating some 1,500 variables accordingly—whether people smoke or would go out with a smoker, fitness, weight, interest in children, and so forth. These are cross-correlated and compared with the variables of others, creating sets of "interactions." Each interaction is scored: a numerical expression of shared trait-tolerance.

"Shared trait-tolerance" is Match's version of YouTube's "viewing time" or Amazon's "Customers who bought this. . . ." item-to-item metric—a similarity measure that offers the greatest insight into future compatibility and companionship. Match, which also owns the OKCupid and Tinder online dating services, has facilitated innumerable relationships. Serious social science research suggests that online dating has transformed how relationships begin and marriages materialize and endure.

A 2013 University of Chicago study found that over a third of US marriages between 2005 and 2012 began online and that online couples typically have longer and self-described happier marriages. "These data suggest that the Internet may be altering the dynamics and outcomes of marriage itself," observed the study's lead author.[23]

The historical point here is not to highlight or speculate on the possible impact or influence of recommenders on marriage. It's to observe that, as recommender systems become more pervasive, they increasingly frame the value and values of what—and who—they recommend. Discussing books, music, movies, and dates divorced from

As recommender systems become more pervasive, they increasingly frame the value and values of what—and who—they recommend.

the recommendations that initially introduced them is becoming more difficult and less common.

Conversely, it's increasingly unusual to enjoy books, music, movies, and dates without turning to recommendation engines for further suggestions. Recommendation, not information, has become a new normal for discovery, serendipity and affirmation.

As machine learning and artificial intelligence algorithms comingle and coevolve with ever-larger and finer grained datasets, recommendations will become smarter, better, wiser, and more persuasive. They'll feel more like "obvious" ahas and epiphanies.

The Google Brain and YouTube coevolution forcefully illustrates this: from 2010, machine learning and artificial intelligence algorithms have successful insinuated themselves as foundational technologies for search and recommendation. Technology journalist Steve Levy described this ongoing transformation in his superb 2016 *Wired* article, "How Google Is Remaking Itself As a 'Machine Learning First' Company."[24]

"Previously, we might use machine learning in a few sub-components of a system," noted Google engineer supremo Jeff Dean. "Now we actually use 'machine learning' to replace entire sets of systems, rather than trying to make a better machine learning model for each of the pieces."

Facebook, Amazon, Alibaba, Microsoft, and, of course, Netflix, have made similar commitments to machine learning. But Google's transparent and transformative commitment to machine learning to redefine its search capabilities into ever-smarter real-time recommendations merits special attention.

For YouTube, Google Brain's "unsupervised learning" has allowed algorithms to find patterns and contexts that utterly eluded formal programming methods. "One of the key things [Brain] does is . . . generalize," YouTube technical lead Jim McFadden told Verge in late 2017. "Whereas before, if I watch this video from a comedian, our recommendations were pretty good at saying, here's another one just like it. But the Google Brain model figures out other comedians who are similar but not exactly the same—even more adjacent relationships. It's able to 'see' patterns that are less obvious."[25]

Brain's algorithm began recommending shorter-length videos for mobile apps and longer videos for YouTube's TV app. Brain correctly—albeit computationally—conjectured that varying video length by platform would result in higher watch times. YouTube launched 190 changes like this that Brain inspired in 2016 and nearly doubled that in 2017. "The reality is, it's a ton of small improvements adding up over time," said Todd Beaupre, group product manager for YouTube's discovery team. "For each improvement, you try 10 things and you launch one." What's more, recommendations are introduced faster than ever. Feedback that once took days to incorporate is now deployed in minutes.[26]

According to the Verge post, more than 70 percent of the viewing time on the site is now driven by YouTube's algorithmic recommendations. The aggregate time people spend watching videos on YouTube's home page has grown twenty times larger than what it was three years before. Brain had completely altered YouTube's "explore vs. exploit" paradigm and practice.

"I had visited YouTube seeking an answer to my question and it had revealed a universe" through Brain-driven recommendation, concluded Casey Newton, who authored the Brain/YouTube post.

Those feelings of discovery and self-discovery are remarkable. But they're exactly what recommendation engines have increasingly been redesigned to evoke. What makes this brief historical overview so striking is not how far or fast recommendation and its technologies have come but how their successes increasingly recall the myths and mystery of the past.

As this is being written, no computer scientist, cognitive psychologist, neurophysiologist, or machine learning maven quite grasps exactly how these Google Brain–type deep learning algorithms work. The underlying processes explaining how this novel suggestion or that serendipitous recommendation emerges from deep learning ensembles remain unclear. In a transcendent irony the Stoics and the ancient Babylonians might surely appreciate, the best recommenders in the world today might almost be described as acts of digital divination. Understanding the matrixed machinations of deep-learning-driven recommendation engines has become a challenge increasingly comparable to understanding the deepest mysteries of the human brain and mind.

HOW RECOMMENDERS WORK

A cynic, Oscar Wilde cynically observed, knows the price of everything and the value of nothing. By contrast, recommendation engines are computational optimists: not only do they know everything's price, they'll also predict its value for you. Successful engines are digitally designed to get to know you better. They're built to learn what you are most likely to like.

More formally, recommenders estimate a "utility function" that automatically and mathematically predicts, ranks, and presents your top preferences. This chapter briefly explains how—and why—that works.

Essentially, recommenders calculate relationships— relationships between people, prices, purchases, preferences, personality, items, images, artists, features, characteristics, metadata, melodies, rankings, ratings, talents, clicks, swipes, taps, tags, texts, locations, moments in time, times of day—that might better inform and improve human choice. These relationships capture and calculate the essential patterns, features, and similarities that characterize and contextualize choice. Recommenders succeed only when users get desirable and actionable choices; they fail if users aren't interested or influenced by the available options.

The recommendations—choices—people ignore can thus prove as predictively revealing as the ones they pursue. Every choice one makes—or declines—becomes useful and usable data. User choices effectively train recommender choices. Recommenders of all kinds from all over the world—Alibaba, Amazon, Booking.com, Facebook, Quora, LinkedIn, Instagram, Netflix, YouTube, Pinterest, Spotify, TikTok, StitchFix—assiduously track and learn from user choices. Their software and systems are dedicated to generating inferences and insights that learn to predict what you might want next in that moment, in that context . . . and possibly beyond.

In that respect, recommendation engines epitomize and extend machine learning's ongoing evolution. Any dataset, algorithm, or line of code that helps a system learn to offer, rank, or optimize choices can be incorporated into a recommender. Recommendation engines don't just recommend; they can and do learn to customize, personalize, and contextualize recommendations. They grow in relevance. Machine learning, in its vast algorithmic variety, has become the primary pathway to smarter personalization and recommendation. Understanding the recommendation engine future now requires understanding the future of machine learning.

Data is key. Better data—about you and for you—make better recommendations probable. Whether that data is discrete and categorical—specific values with no scale attached; yes/no; present/absent; red/blue/green—or continuous, with natural, numeric scales—temperature, time, height, weight, or cost. Different and diverse data-types feed and fuel the algorithmic ensembles that classify, cluster, rank, and predict recommender relationships. These ensembles define and determine the contextually relevant similarities and surprises that rank and deliver appealing recommendations.

Three core data sources drive recommender design: *users*, *items*, and the *interactions* between them. Users, of course, are the people looking for videos to watch, music to hear, stories to read, foods to eat, friends to meet, emojis to text, advice to follow, and products and services to buy. Recommendations are rated, ranked, and optimized for them. User data is all about profiles, preferences, and personalization. User features, characteristics and choices must be captured, sorted, and quantified.

Items are the particular objects, subjects, and experiences–videos, songs, stories, books, travel, dining, jobs, images, clothing—that recommenders recommend.

Item data describe qualities, features, and attributes that make those items distinctive or desirable; action and/or comic book or romcom movies, for example; sad, up tempo, or hit songs; luxury, business, and/or downtown hotels. How might items be classified, categorized, labeled, or tagged to enhance reliable recommendation? That's how items can become more than the sum—or product—of their individual elements.

Interaction happens whenever users meet items. The perceived value—or utility—of those interactions determine what gets recommended and why. Is there a match? Was there a click, a swipe, a purchase? Interactions are typically characterized as either *explicit* or *implicit*. Explicit interactions reflect deliberate user decision to rate, rank, or review items. Facebook's "thumbs-up" likes and Amazon's stars and reviews, for example, explicitly signal perceived utility. Users unambiguously declare a preference or affinity.

Implicit interactions, by contrast, capture user behavior. As the YouTube case illustrates, technically tracking click-throughs is simple, as is measuring time spent on videos viewed. Onscreen interactions—swipes, taps, strokes, and/or dwell-times—can be recorded to draw meaningful data-driven inferences about user preferences, that is, people "like" videos they view for longer times. The relative ease of digitally instrumenting implicit behaviors explains why data scientists worldwide have come to consider them more valuable for recommendation than explicit ratings. Statistically speaking, (inter)actions speak louder than ratings.

Some "implicits" may initially mislead; for example, they might playlist songs you don't really listen to. With time and training, though, recommendation engines learn which behaviors predictively reveal personal preferences. Increasingly, recommenders are explicitly designed and built with implicitness in mind.

Collating the features, elements, and attributes of users, items, and interactions creates the datasets for recommendation. The more data—and metadata—Netflix has about its videos, viewers, and views; the more details Amazon commands about its customers and its products; the more interaction and engagement Facebook facilitates between families, "friends," and "followers," the more precise and persuasive their recommendations are likely to be. No Wildean cynicism here: successful recommendation engines prioritize personalization over price. They get greater value from users by learning what users value.

Algorithms transform data into relevant recommendations by finding, calculating, and ranking the most interesting correlations and co-occurrences for users. Recommender algorithms typically run two essential computations on the user-item-interaction nexus: prediction and selection; that is, rating prediction and item selection.

Algorithms transform
data into relevant
recommendations by
finding, calculating,
and ranking the most
interesting correlations
and co-occurrences
for users.

Ratings prediction assigns scores (ratings) to potential items that reflect known user preferences. Item selection, by contrast, combines ratings with other data-driven criteria to sort and/or rank item recommendations. For example, context—time of day or location—could strongly influence which items to suggest; a nearby restaurant for lunch or store for shopping.

The more recommendation engines are used, the more reliably they marry, merge, and mash-up prediction and selection algorithm outcomes. The proffered recommendations themselves become data sources. The most successful engines power virtuous learning cycles. They feed and fuel themselves on the very data they help generate.

Every speck of explicit and/or implicit data collected and processed for prediction and selection is quantified. As Christian Rudder, a cofounder of pioneering dating recommender site OKCupid, notes in his *Dataclysm*:

> Algorithms don't work well with things that aren't numbers so when you want a computer to understand an idea, you have to convert as much as you can of it into digits. . . . The challenge facing sites and apps [and recommenders] is thus to chop and jam the continuum of human experience into little buckets 1, 2, 3 without anyone noticing: to divide some vast, ineffable process—for Facebook, friendship, for Reddit, community, for dating sites,

love—into pieces a server can handle. At the same time, you have to retain as much of the *je ne sais quois* of the experience as you can so the users believe what you are offering represents real life.[1]

Finding that balance requires humanizing quantification and quantifying humans. Explicitly asking people to rate movies or tacitly tracking the time users take online reading reviews delivers hard numbers. Age, gender, location, time of day, and shoe size are also easy to quantify. Quantifying current user preference enables computational inference and insight into future user preference.

Generally, recommender algorithms classify or run regressions on quantified data. Classification algorithms predict in which class or category a data point belongs—for example, male/female; luxury, business, family; cheap, moderate, expensive; colors, customer sentiment.

Regression algorithms, in mathematical contrast, use data to predict continuous outcomes rather than sort into discrete categories. That is, the answers are numerical quantities like "propensity to buy" or "number of minutes," not allocation into specific groups.

So where classification is used to determine whether or not it will rain tomorrow, regression would predict the likely amount of rainfall. In recommendation engine contexts, a classifier would determine what songs you might play and regression would predict how long you might play them. Classification + regression = predictive personalization.

Comparing Similarity

That predictive personalization sensibility makes 'similarity' recommendation's secret sauce. Whether broadly or narrowly defined, similarities explore, exploit, and explain the data-driven dynamics of "people like you" and "people who bought '*this*' also bought '*that*'" recommendation and suggestion. Seemingly obscure data patterns burst into compelling relevance. What common elements—what features—do desirable items possess? What user aspects and attributes imply comparable tastes and preferences? What clusters and classes of similarity are most likely to matter most? Algorithms—individually and collectively (known as "ensembles")—compute those answers.

Ironically but importantly, similarities can be sliced, diced, and digitally defined along radically different dimensions. "Different similarities" can range from the virtually identical—a (slightly) different hue of shoe that somehow makes it visually *pop*—to oblique but profound commonalities that command attention—for example, that quirky but charismatic character actor who always makes a movie memorable, or the music producer whose acoustic audacity creates an infestation of earworms.

Dating recommendation engines, for example, take remarkably dissimilar approaches to similarity. Tinder and eHarmony bring profoundly incompatible optimization perspectives to matchmaking recommendation. "Attractiveness" for a fling or hook-up might be calculated very differently than for committed "let's grow old together" relationships.

Quantifying quirky similarities makes ever-more-personalized and provocative recommendation possible. That said, while improving predictive power is wonderful, predictability itself is not. Surprise matters enormously. Variety of similarity is, indeed, the spice of recommendation life. Mathematical similarities of variety—and the mathematical variety of similarities—enable recommendation engines to escape the predictability trap. Given timely and diverse data, recommenders learn to discriminate between subtle and significant change.

The paradoxical genius of similarity is the mathematical promise of predictable surprise, also known as "serendipity." That's how recommenders successfully program music playlists that energize your day, suggest restaurants to explore new cuisines, and introduce you to friends you haven't met yet. They computationally leverage the essential insight that, "at their core, recommendation systems are nothing but similarity hunters."[2] The calculated likelihood of surprise—a timely and contextually relevant surprise, to be sure—turns them into "serendipity hunters," as well. If recommendation engines don't consistently make their users feel lucky, something's gone wrong. Successful recommender algorithms artfully navigate between continuity and novelty.

What do the "similarity hunting grounds" of recommendation actually look like? "Here is the cold, hard truth," declare recommendation pioneers Joseph Konstan and John Riedl. "You are a very long row of numbers in a very, very large table. This row describes everything you've looked at, everything you've clicked on, and everything you've purchased on the site; the rest of the table represents the millions of other Amazon shoppers. Your row changes every time you enter the site, and it changes again with every action you take while you're there. That information in turn affects what you see on each page you visit and what e-mail and special offers you receive from the company."[3]

Don't limit those cold, hard truths to Amazon; they define and describe recommender architectures worldwide. Lengthy rows and enormous columns are much more than humongous data structures; they're multidimensional platforms for dynamic mass-customization of meaningful similarities. How do they do it?

Three overarching design themes predominate: content-based systems that rely on the properties and characteristics of items; collaborative filtering systems that recommend items based on similarity measures computed between users; and hybrid systems that ensemble the best aspects and elements of content and collaborative filtering systems to produce recommendations superior to either one alone. Their underlying mathematics quickly grows complex but let's start with the simplest recommender approaches first.

Most Popular

Popularity is recommendation's simplest success story. Display the most popular items along some distinctive dimension: the most popular movies, news stories, destinations, or clothing. What items are most popular right now? Present the week's best sellers. Display the stories that are most read/most shared/most tweeted. What's most popular for men? Women? Popularity is the transcendental similarity. Sales transactions and systems logs deliver the necessary data; no complicated math necessary. But why not make popularity the gateway recommendation to personalization and profiles? Track which items win the most swipes and sign-ups. Monitor users, items, and interactions. Popularity is the most popular way to recommend; it's the easiest and most common way to deal with the "cold start" recommender problem.

Association Rules and Market Basket Models

Association rules and "market basket" analyses are fraternal quantitative twins; they use the same statistical calculations to find items purchased together. They mine historical transactions data to identify frequent co-occurrences unlikely to have happened by chance. In other words, *what are the odds these items would be purchased together?* Back in the 1990s, for example, Walmart discovered customers who bought Barbie dolls had a 60 percent likelihood of also buying one of three types of candy bars. The canonical—if apocryphal—"association" vignette has new fathers concurrently buying beer and diapers when they shop. Draw your own co-occurrence inferences.

When customers acquire items one at a time (a bank loan, for example) that's an association. When they buy several items at once, that's a market basket. "Association analysis" is calculated at the customer level—to wit, "what's in their account?" "Market basket analyses" happened at the point of transaction—"what's in their basket?"

"Association rules/market basket recommenders" require three steps:

1. Calculate the sales/purchase relationship between each item offering and every other item offering using "association math" statistics. That is, compute all possible co-occurrences. This creates a rather large spreadsheet of sales co-occurences.

2. Identify and prioritize the pairings most co-related with boosting sales and/or margins. Customers with combined credit and debit cards, for example, might be three or four times more likely to hold an auto loan than a randomly selected auto loan customer.

3. Play the probabilistic odds by personalizing offers for customers who have one item of strongly associated pairs but not the other.

Simple and computationally swift, co-occurrence recommenders require minimal data prep and processing. Detailed customer knowledge beyond existing products is unnecessary. (Good for privacy.)

Firms with limited offerings find association and/or market basket analytics make personalization computationally tractable. But "caveat vendor": this approach inherently emphasizes sales over customer satisfaction. The analytics are transactional, not interactional. These analytics are similar to Amazon's "items bought together" recommender success. That's no accident.

Content Filtering, Collaborative Filtering, and Hybrids

Mathematically, these recommendation engines live in the matrix. More specifically, they live in the utility matrix.

The enormous tables or spreadsheets Riedl and Konstan discuss above are matrixes that hold the affinities and preferences—the data-defined relationships—between users and items. Each row and column contains essential data about the relevant characteristics and attributes of users and items alike—users are rows; items are columns.

Each user/item intersection—each matrix pairing—reveals the degree of user preference for that item. For content-based and collaborative filtering recommenders alike, the utility matrix is where the search for similarities pays off. But not every user will have stated or disclosed an item preference or ranking. Most items, in fact, won't be formally rated at all. In technical terms, those gaps make the matrix "sparse," that is, more empty than full—like a computationally enormous Swiss cheese. That's the technical and mathematical problem to solve.

The recommender goal and aspiration is to reliably predict—fill in—those utility matrix blanks. Recommenders use algorithmic tricks and techniques to infer and impute those missing user preferences from existing data. Whether shopping on Amazon, binge-watching Netflix video, or listening to Spotify playlists, matrix mathematics—the mathematical manipulation of those lengthy rows and massive columns—focuses on personalizing the predictions of what people will most likely like.

Content-based and collaborative filtering recommenders draw upon different types of similarity to populate their respective utility matrixes. Content-based recommenders rely on the assumption that items with similar properties and features will be rated similarly. Determining which features and attributes are most predictive for a given user is the challenge. Collaborative filtering's core assumption is that people with similar tastes will rate things similarly. That is, "people like you" who've enjoyed *this* movie, restaurant, or blog post will most likely like *that* one. Collaborative filters consequently–even ironically—needn't know or understand anything about the items they recommend.

Content-based recommenders need to create both item profiles and a user profile—records or logs that capture and represent the most important features and attributes of that item or user. An item profile is a matrix of "items to features" describing key item attributes—actors in movies, say, or song lengths or types of restaurant cuisines.

User profiles rely on the same features as the constructed item profiles. But here the matrix's rows represent the user. Those values represent the degree of user affinity or preference for the feature. Recommenders then compute the similarities between the item profiles and the user profiles. The closer those calculated similarities, the more likely the recommendation makes sense.

So item profiles for movies might list key features such as actors, directors, genre, budget, or Academy Award winner. A particular feature's presence or absence

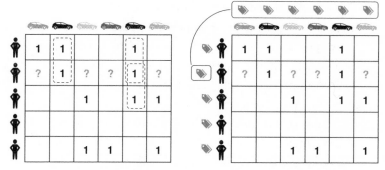

Collaborative: *similar interactions*   Content-based: *similar features*

Figure 1 Adapted from Marcel Kurovski, "Deep Learning for Recommender Systems," https://ebaytech.berlin/deep-learning-for-recommender-systems-48c786a20e1a

might be demarked by a 1 or a 0. The user profile captures whether—or how much—users "like" the presence (or absence) of particular features. Content-based recommenders build user profiles with a set of weighted features. Those weights might be a Boolean (0 or 1) capturing, for example, whether the user liked/disliked that feature. As OKCupid's Rudder observed, features could have other numerical metrics—or weights—associated with them.[4]

To grossly but usefully oversimplify, if a user watched ten movies, one could reasonably infer affinity or preference by calculating which features (actors, directors, genres) were most often present—or absent—in those films. If six of those chosen movies were big-budget action films featuring famous male stars, for example, the user's preference/affinity profile and utility matrix would be radically different from users who viewed four big budget romcoms, two small independent foreign films, and an Academy Award–winning documentary.

Finding the key features—or features cluster(s)—positively correlating to user preference would likely prove easier for the former viewer than the latter.

How do specific recommendations computationally emerge from the data? They come from identifying, defining, and inferring desirable similarities. Consider the features and attributes that go into defining user and item profiles. You have rows; you have columns. You have numerical values filling those rows and columns.

Most simply: calculate correlations between user/item datasets to predict recommendations. Greater correlation implies greater similarity. (Note that "correlation doesn't imply causality." That is, there is no attempt to explain an underlying "why" for that correlation.)

Correlations are commonly and easily calculated but come with limitations and constraints (as do all such statistical techniques.)

Alternately, think of those rows and columns as sets of numbers. You'll recall a set in mathematics is simply a well-defined collection of distinct elements. To what extent does the set of features a user "likes" (or are present in items that user likes) overlap with the feature set an item—such as a movie, a book, a song, a restaurant, or an investment—actually possesses?

That overlap—the intersection of those sets—is another way of mathematically describing their similarity. To wit, the more similar the sets, the more likely users will like those items. Some recommenders compute overlap between "user profile" sets and "item profile" sets. Even if the number and size of these sets is enormous, computers can quickly and easily calculate and rank their similarities. The denser and richer the item/user overlaps, the more likely those items will be recommended.

These rows and columns can be mathematically described in other powerful and useful ways. Instead of seeing them as sets, start viewing them as vectors. Recall that a vector is a mathematical object that has direction and magnitude. Wind, for example, has speed (magnitude) and a direction and is often mathematically modeled with vectors.

Recommendation engines use "feature vectors"— vectors containing multiple elements about an object or item—for their work. Feature vectors are ordered lists of quantified attributes that live in multiple dimensions. In other words, they're spatial sequences of numerical data relevant for prediction or preference.

Combining feature vectors of items and objects together creates feature space. This multidimensional feature space is where patterns and geometries of similarity are found. Feature vector representations are commonly used in machine learning and recommenders because they so readily lend themselves to mathematical comparison.

Remember the Pythagorean theorem, $a^2 + b^2 = c^2$? One simple way to compare the feature vectors of two items is to calculate their Euclidean distance. Just how far apart are these vectors? The closer they are, the more similar they are. An alternative—or complementary—trigonometric approach: calculate the angle between vectors. The smaller the angle—that is, the closer the angle to zero—the more similar those vectors are. (Being perpendicular or "orthogonal" implies maximal dissimilarity.) This technique is known as "cosine similarity."

So for movies, for example, a recommender would "vectorize" the weighted features, attributes and descriptions, such as budgets, genre, awards, and reviews. These feature vectors would digitally encapsulate the movies' item profile. The user profile would consist of a feature vector aggregating the feature vectors of movies the user had liked. A combination of Euclidean distance and cosine similarity, for example, could be used to filter, rank, and recommend the next most likable movies to see.

For comparing blogs and news stories one could alternately use "bag of words" or term frequency-inverse document frequency (TF-IDF) algorithms to vectorize their contents. That is, identify the features—the words, names, and phrases—to build document profiles for comparison.

Those TF-IDF–defined vectors contain the word counts of their respective documents. (TF-IDF is a statistical measure that evaluates how important a particular word is to a document in a collection or "corpus."). Cosine similarity measures the angle between two vectors projected into a multidimensional space where each dimension corresponds to a word in the document. This captures the relative spatial orientation of the documents, Again, the smaller the angle between vectors, the more similar the documents.

But understanding precise computational details is less important than the broader conceptual point: recommendations emerge from innovative computational efforts to locate, identify, and predict relevant similarities between data. Think of a vast cloud—even galaxies—of data points. The greater the calculated data distances, the less similarity; the closer the data distances, the greater the similarity. Distance is the proxy for similarity.

Feature space, for content-based recommenders, is the happy similarity/serendipity hunting ground. Collaborative filtering recommendation engines use similarities to track recommendations differently. Item content matters not at all; the similarities between users—between people—are what get measured and what matters most.

Think of a vast cloud—
even galaxies—of
data points. The greater
the calculated data
distances, the less
similarity; the closer
the data distances, the
more similarity.

User-based recommenders thus embrace a neighborhood and community sensibility: gather groups of people who have rated items, movies, music, restaurants, or dates; calculate the mathematical distances—the (dis) similarities—between each "people pairing" based on their ratings; find those individuals most similar/closest to you; recommend items to you that "people like you" also liked. These ratings—those data—can be implicit, explicit, or both.

Again, user similarities—just like content similarities—can be computed. Assign a similarity weight to all users relative to the active user; select users with the highest similarity to the active user—also known as "the neighborhood"; and then compute predictions from a weighted combination of the selected neighbors' ratings. Quantitatively, this mathematical neighborhood is filled with people like you.

While user-based collaborative filtering demonstrably works well, creating, maintaining, and reliably updating the neighborhoods can be computationally expensive. Just how big should these neighborhoods be? With millions—even hundreds of millions—of users, calculating the "right sized" neighborhoods is computationally expensive and inefficient.

So look for clever shortcuts. Instead of statistically sorting through neighborhoods of similar users, computationally create item-based neighborhoods. That is, look for items similar to items the user has already rated highly, and recommend the most similar items. This is what Amazon has done so successfully for so long. Does this echo the association rules recommendation approach described earlier? Yes. Recommendation engines computationally evolve.

Note that item-based collaborative filtering doesn't rely on item features for similarity hunting. Similar items are found and discovered in item neighborhoods based on user behaviors—ratings, reviews, and dwell time, for example.

Collaborative filtering's data-intensive demands typically mean that item-based rather than user-based computations scale better and more efficiently. But the essential computational purpose remains: predicting future interactions based on the past. How best to do it? Two method types are typically used: memory based and model based.

Memory-based techniques largely rely on straightforward similarity measures. Remember the matrix: if there's a huge matrix with users on one side and items on the other with their cells containing likes or ratings, then memory-based techniques use similarity measures on two vectors (rows or columns) in the matrix to determine similarity.

Memory methods have problems typically associated with large "sparse" matrixes; the number—the density—of user-item interactions is frequently too low to generate high-quality/quantity clusters and neighborhoods.

Model-based methods are more computationally ambitious. Using machine learning and data mining techniques, they "guess" how much a user will like an item that they've not seen before. The aspiration is to train predictive models that learn. By training on item vectors for specific users, for example, models predicting personalized user ratings for newly added items become possible. Existing user-item interactions could, say, train a model to reliably predict, say, the top five items a user might most like. Technically, model-based methods can computationally recommend larger number of items to larger number of users versus memory-based methods. This gives them larger "coverage," especially with large sparse matrixes.

Essentially, memory-based methods use all the known data all the time to make their predictions while model-based algorithms use data to learn to be better predictors. Machine learning infuses a dynamism that memory cannot. In theory and practice, this critical distinction makes model-based methods the better bet for the recommender future.

This multiplicity of methods highlights the real importance of hybrid recommenders. Content and collaborative filtering recommenders each have their challenges and strengths.

As the Netflix Prize competition revealed, hybrid recommenders that combine—ensemble—complementary content and collaborative filtering systems produced recommendations superior to either approach alone. Mixing, matching, and mashing up different recommender algorithms typically outperforms any single method. This discovery has forced recommendation engine designers, architects, and engineers worldwide to take more holistic, integrated, and interoperable approaches to personalization. While adding undeniable complexity, hybrids consistently deliver high-performance results. Hybrid examples include

Weighted   Implement different methods separately and then combine their predictions

Feature Combination   Features from different recommender systems data sources are put into a single recommendation algorithm

Feature Augmentation   The output of one system is used as an input feature for another; for example, features generated by one model-based method is used as input for another model.

Cascading   One recommender system refines the results of another

Meta Level   A model learned by one recommender is used as input for another. The difference from feature augmentation is that the entire model becomes an input

Mixed   Incorporates two or more techniques at the same time, for example, combines content-based and collaborative filtering

Of course, other ways of ensembling and interconnecting recommender methods exist. But the deeper insight

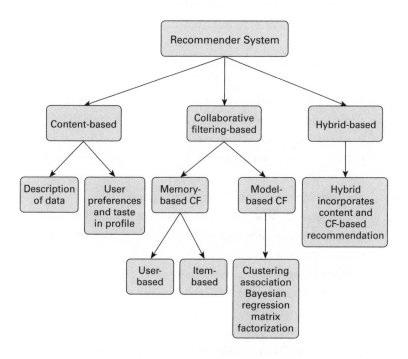

Figure 2   Adapted from H. Mohana and M. Suriakala, "A Study on Ontology Based Collaborative Filtering Recommendation Algorithms in E-Commerce Applications," *IOSR Journal of Computer Engineering* 19, no. 4 (2017): 14–19.

is that recommendation engines shouldn't be designed or developed as monoliths. To the contrary, increasing model diversity may be more important than improving individual models. (Indeed, the Netflix Prize–winning ensembles contained over one hundred models.) Netflix, of course, offers a very successful, very effective hybrid recommender system that is always learning.

Dimensionality's Curse and Latent Factor Insight

Another key insight from the Netflix Prize came from confronting the enormous size, sparsity, and multidimensional complexities of the matrixes themselves. The collaborative filtering matrix was simply too big, too messy, and too complicated; a fundamental rethinking was required.

Sparsity and scalability have proven the biggest challenges. Collaborative recommendation engine performance dramatically degrades as the numbers of data dimensions rise. This so-called "curse of dimensionality" undermines the ability to get more value from more data.

The seemingly counterintuitive performance and predictability breakthrough came from deconstructing the user-item matrix. Less data can mathematically yield more information. "Dimensionality reduction" is key: get more recommendation information and insight by computationally cracking the giant matrix into smaller matrixes. In other words, *find lower-dimensional representations of the data that retain as much information as possible*. For recommendation engines, singular value decomposition (SVD) is a "matrix factorization" technique that creates the essential submatrixes. SVD is a method for deconstructing and describing a matrix into three parts:

$A = USV'$

Where U and V are orthogonal and and S is diagonal. Think of this deconstruction as the rows U reveal how much each user likes a particular "feature" while the columns of V tell us which items have each feature. The diagonal of S "weights" and details the overall importance of each feature. What exactly are these features? We don't know; they're what the SVD has determined from the data.

To get a quick sense of how this works, picture that these features are actors, genres, budgets, directors, length, prizes, and so on, for a movie. Different movies would have different mixtures of different features. Different viewers would weight these different features differently. What makes SVD explicitly intriguing is we don't have to decide or determine what those features are. The SVD does it for us.

Mathematically deconstructing—or dividing—the larger matrix into these smaller matrixes surfaces hidden or "latent" features that measurably influence user ratings.

That is, these features weren't directly observable until the big matrix was broken into the product of the littler matrixes. Dimensionality reduction and decomposition reveal and highlight data-based relationships otherwise concealed in the larger matrix. By definition, these smaller matrixes are less sparse than their parent.

These lower-dimensional matrixes offer intriguing intuitive rationales for inferring tacit user preference. To wit, a user gives good movie ratings to *Inception*, *Edge of Tomorrow*, and *Arrival*. These aren't presented as three distinct opinions but suggest a strong predilection for high concept "time-twisting" science fiction films. Unlike more specific features—such as actors, budgets, and genre— latent features are expressed by higher-level attributes. Matrix factorization is a powerful mathematical mechanism offering inferential insight into how aligned users may be with latent features and how comfortably movies fit into latent feature sets.

This "similarity" is qualitatively and quantitatively different—and conceptually deeper—than standard nearest neighborhood calculations. Even if two users haven't rated any of these same movies similarity discovery may still happen if, as these latent features suggest, they share comparable underlying tastes. SVD and other dimensionality reduction approaches such as Principal Component Analysis calculate new dimensions of predictive similarity for recommendation.

Machine Learning, Bandits, and Explanation

Bluntly, no grand unified theory or universal algorithm for recommendation exists. Instead, there are staggering arrays—huge n-dimensional matrixes—of interrelated recommender design opportunities, options, and approaches. Nearly every invention and innovation from data science, statistics, and machine learning ultimately finds its way into some recommender system instantiation. Conversely, virtually every notable recommender improvement draws implementational inspiration and insight from statistics, predictive analytics and data science innovation. That will not change.

As discussed, the most exciting and influential advances in recommender system capabilities come from algorithmic advances in machine learning.

To the extent past is prologue, the future of machine learning is the future of recommendation engineering. That said, it is no less true to observe that the future of recommendation engineering is also the future of machine learning. Recommenders aren't merely or simply a special case of machine learning; they uniquely drive technical and commercial machine learning research agendas worldwide. Epistemologically, ontologically, teleologically, rhetorically, psychologically, socially, and economically, recommendation is one hell of a machine learning problem to solve.

Machine learning can be "supervised," where the algorithm takes labeled data and creates a model that can make predictions given new data (again, classification or regression); the learning can be "unsupervised," where data are not labeled or categorized. The goal then is algorithmically finding patterns or imposing structure in order to derive meaning (clustering and dimensionality reduction described above are two examples). Learning can also be reinforced where reward systems and trial-and-error experimentation combine to maximize long-term reward.

While technical details describing machine learning's transcendent impact on recommenders goes beyond this explanatory chapter's scope, highlighting the essential knowledge of its underlying principles does not.

To be sure, these core principles need to be understood in the context of similarity, serendipity, discovery, and relevance. But they represent the surest way to begin learning how machines learn to recommend.

Data has shape—whether in the form of matrixes, graphs, trees, or neural networks—and shape has meaning. How can—how should—machine learning make those shapes more meaningful—or more meaningfully effective—for recommendation? That question requires an ensemble answer.

In his terrific essay, "A Few Useful Things to Know about Machine Learning," University of Washington's Pedro Domingos, a leading ML researcher (who also wrote

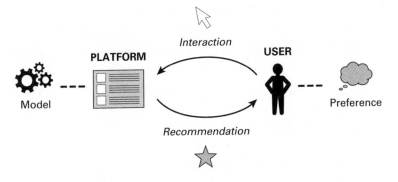

Figure 3 Adapted from Allison J. B. Chaney, Brandon M. Stewart, and Barbara E. Engelhardt, "How Algorithmic Confounding in Recommendation Systems Increases Homogeneity and Decreases Utility," *Proceedings of the 12th ACM Conference on Recommender Systems* (2018): 224–232.

The Master Algorithm) structures his discussion in a manner that maps beautifully to the recommender problem. He describes his "machine learning equation" as follows:

Learning = Representation + Evaluation + Optimization[5]

Note well: every recommendation engine approach presented thus far—content or collaborative; model or memory—could be distilled into this formula. Understanding how recommenders personalize and prioritize recommendations requires understanding how representation, evaluation, and optimization collectively define learning. Domingo's equation, pragmatically interpreted by Harvard's Joe Davison, is paraphrased and elaborated upon below. Davison takes special pains to distinguish "doing statistics" from "machine learning."[6] The former is about computation; the latter is about transforming computation.

Representation involves the transformation of data from one space to another more useful space where interpretation becomes easier. "Raw pixels are not useful for distinguishing dogs from cats, so we transform them to a more useful representation which can be interpreted and evaluated," notes Davison. What feature spaces deliver the most tractable information?

Evaluation essentially determines the accuracy and reliability of your classifier; that is, what's the "loss function"? (What was lost in translation? Where are the gaps?) Did your algorithm effectively transform your data to a more useful space? Did you correctly predict the next word in the unrolled text sequence? These questions tell you how well your representation function is working; more importantly, say Davison and Domingo, they define *what it will learn to do*. That is, how one chooses to evaluate learning determines what learning actually and practically means. Does the recommendation engine learn to maximize click-throughs or increase dwell times? Is the recommender learning to recommend what you are most likely to like? Or is it learning to recommend the most profitable items you are most likely to buy? These distinctions aren't subtle; they're essential to learning what is being learned.

Optimization is the final piece of this puzzle. With an evaluation component, one can *optimize* the *representation function* in order to improve one's *evaluation metric*. That is, evaluation criteria shape what representational features and elements need to be optimized. One must optimize how one optimizes.

"The choice of optimization technique is key to the efficiency of the learner," Domingo has observed. "It is common for new learners to start out using off-the-shelf optimizers, which are later replaced by custom-designed ones."

This triad clarifies that recommendations are as much vital ingredients as the products of machine learning. Whether that learning is largely supervised, unsupervised, or embraces ensembled neural nets matters far less than the question of what—and how much—the recommendation engine truly seeks to learn.

Consider, for example, "learn to rank" algorithms. Traditional machine learning solves prediction problems (classification or regression) one at a time. With email spam detection, for example, you'll identify all the features associated with an email and classify it as spam or not. Traditional ML creates a "class" (spam/not-spam) or a single numerical score for that instance.

Learn to rank, by contrast, solves a ranking problem for a list of items. Learn to rank's goal and purpose is to optimally sequence those items. So "learn to rank" doesn't care about individual item scores as much as the relative ordering of the items.

For recommendation engines, "learn to rank" means that individual recommendations matter less than the ordered set of recommendations. That is, the recommender is less interested in learning what recommendation to give users than what set of recommendations to present. That distinction—for users and machine learning alike—is not subtle. If users respond better and more reliably to ranked recommendation sets over individual recommendations, that's well worth knowing. These kinds of design decisions increasingly shape recommendation engine investment and development.

But how should those decisions be determined? What criteria should be used to prioritize which recommendations are generated? Netflix, Google, TikTok, Stitch Fix, Pinterest, LinkedIn, and other recommender-driven organizations take a deliberately experimental approach to those questions. They want to learn when it makes more sense—and more money—to explore novel recommendation options versus exploiting existing ones.

This goes beyond simple A/B tests with particular recommendations. Think instead of virtual slot machines where recommendations become arms of the bandit. Multiarmed bandit algorithms seek to maximize the reward taken by a decision; for example, maximizing click-through rates or watch times of personalized shows or movies. Multiarmed bandit algorithms represent a form of reinforcement learning. Reinforcement learning looks at the actions software agents should take to maximize some defined notion of cumulative reward.

The popular epsilon-greedy bandit algorithm, for example, continuously balances exploration with exploitation. In greedy experiments, the lever with the highest known payout always gets pulled except for random intervals. Then a randomly chosen arm (recommendation) is pulled a fraction of the time. The rest of the time, the highest-returning arm gets pulled.

"Experiments based on multiarmed bandits are typically much more efficient than 'classical' A-B experiments based on statistical-hypothesis testing," states Google's Steven Scott. "They're just as statistically valid and . . . can produce answers far more quickly. They're more efficient because they move traffic towards winning variations gradually instead of forcing you to wait for a 'final answer.' . . . Basically, bandits make experiments more efficient so you can try more of them."[7] That's important; it guarantees greater learning.

Again, recommendations become essential data for training smart algorithms. At the same time, organizations and their machines can learn what recommendations might be worth more to explore than to exploit.

But amid this machine learning exploration, exploitation, and experimentation explosion, a fundamental problem grows exponentially worse. The increasing size, sophistication, and complexity of recommender systems renders them opaque. They offer seemingly superior and serendipitous suggestions but cannot explain themselves. The "why" has vanished.

"Deep learning systems are huge, deep and inaccessible—they operate like black boxes," declares Imperial College's Alessio Louscio.[8] The very qualities that give deep learning such remarkable predictive power subvert its ability to accessibly and transparentlycommunicate its reasons and rationales. The underlying "latent factors" may be too latent.

"Machine learning explainability" has consequently become one of the most important areas in recommender systems research. Serendipitous recommendations aren't sufficient; people need to be able to see and grasp the underlying "why."

As one researcher observes, "Systems that cannot explain their reasoning to end-users risk losing trust with users and failing to achieve acceptance. Users demand interfaces that afford them insights into internal workings, allowing them to build appropriate mental models and calibrated trust."[9]

In other words, explainability has become a recommendation engine essential. Indeed, one research group ran an experiment incorporating explainability as a weight for ranking travel recommendations. They found users typically more satisfied with explainable recommendations. Spotify's Paul Lamere coined the neologism "recsplanation" to describe how recommendation engines justified their offers.[10]

Perhaps unsurprisingly, demands for explainability and interpretability in machine learning systems are increasingly being met by, yes, machine learning systems. The very same algorithms that create relevant recommendations are being used to classify, categorize, and cluster explanations for them.

In his usefully provocative "The Mythos of Model Interpretability," Carnegie Mellon's Zachary Lipton asks, 'What do we mean by interpreting a machine learning model, and why do we need it? Is it to trust the model? Or try to find causal relationships in the analyzed phenomenon? Or to visualize it?"[11]

These questions—which have moved from rhetorical to practical shockingly fast—highlight a fantastic meta-irony: the ultimate destiny of intelligent recommendation engine design will be recommendations that can credibly and persuasively learn to recommend themselves to their users. They will use all the techniques and technologies above to better learn about themselves so that they can make better recommendations to their users.

EXPERIENCING
RECOMMENDATIONS

Everything is a recommendation. Full marks to Netflix's data scientists for their bravura design principle and insight. They should know. As important as relevance and reliability might be, how users actually experience recommendations matters more. Are they curious? Energized? Motivated? Decisive? Bored? When everything is a recommendation, recommendations are everything. What should the RX—recommendation experience—look, sound, and feel like? This question dominates the high-performance recommendation engine design future.

The experiential evolution of recommenders has been astonishingly fast. What began as information retrieval techniques for document search swiftly transformed into sales tools, trusted advisors, and immersive environments as the internet commercialized. At virtually every touch- and-turning point, recommendation has become more integral to enhancing user experience online. This is as true for movies, music, and shopping as for travel, dining and socializing. In many cases, recommenders become indistinguishable from user experience; indeed, getting great recommendations may be the entire point and purpose of the interaction.

People from Boston to Beijing to Berlin to Buenos Aires expect good recommendations. They expect relevant choice. User experience is tightly coupled with user expectations; you can't change one without affecting the other. People expect personalized offers that accurately reflect—and even anticipate—their preferences. They want their personalization simple, easy, and now. Rising user expectations alter recommendation design trajectories. People don't just demand better recommendations; they want better recommendation experiences. It's the difference between good cooking and fine dining. Great meals offer more than great food—the ambiance, the service, the conversation all enhance both the food and the experience. Recommendation engines draw upon multidisciplinary knowledge and expertise to deliver greater value and inspire greater user confidence.

That said, recommendation engines always power their own agendas: they alternately seek to inform, advise, anticipate, persuade, nudge, seduce, share, upsell, cross-sell, build trust, increase "stickiness," inspire loyalty, delight, surprise, elicit ratings and reviews, create community, and personalize, personalize, personalize. And yet, they don't want to appear creepy, intrusive, and/ or exploitive. They must serve the needs of enterprise and user alike. Accommodating those tensions, conflicts, and aspirations is what user experience design and expectations management is all about.

Recommendation engine designers consequently look to cognitive and social psychology, sociology, behavioral economics, and other research domains for information and inspiration. They want to better understand what makes recommendations work for users and themselves. New technologies create new contexts for new recommendations. New recommendations, in turn, transform customer behavior,

Netflix's transformation highlights the challenge of creatively aligning user experience with user expectations when "everything is a recommendation." The company's fervent commitment to personalization transcends technology and device. "We have discovered through the years that there is tremendous value to our subscribers in incorporating recommendations to personalize as much of Netflix as possible," observed data scientists and engineers Xavier Amatriain and Justin Basilico.[1]

Netflix leadership understands that future success depends as much on the quality of Netflix RX as the quality of its content. Recommendation innovation must measurably impact and influence user behavior. But who would have thought recommender systems originally born to manage DVD inventory would grow up to breed binge viewers?

Barely a year after announcing its Netflix Prize recommender competition, the company launched its web-streaming service. That 2007 move—the same year that Apple's iPhone launched—changed everything. Apple's iPad, introduced three years later, both accelerated and intensified that change. Marrying touchscreens and streaming would rewrite every rule of recommender experience design.

"Streaming has not only changed the way our members interact with the service," Netflix declared in a 2012 techblog, "but also the type of data available to use in our algorithms. When we were a DVD-by-mail company and people gave us a rating, they were expressing a thought process. You added something to your queue because you wanted to watch it a few days later; there was a cost in your decision and a delayed reward. With instant streaming, you start playing something, you don't like it, you just switch. Users don't really perceive the benefit of giving explicit feedback, so they invest less effort."[2]

Replatforming recommendation around streaming forced a fundamental algorithmic and experiential rethink. "We adapted our personalization algorithms to this new scenario in such a way that now 75% of what people watch is from some sort of recommendation," recalled Amatriain and Basilico. "We reached this point by continuously optimizing the member experience and have measured significant gains in member satisfaction whenever we improved the personalization for our members."[3]

Netflix not only personalized subscriber home pages with data-driven rows of customized recommendations, it subtly prompted users to pay closer attention to the process. "We want members to be aware of how we are adapting to their tastes," noted the Netflix engineers. "This not only promotes trust in the system, but encourages members to give feedback that will result in better recommendations."

Increased awareness was but a first step; Netflix wanted recommendation to encourage understanding, as well. "A different way of promoting trust with the personalization component is to provide *explanations* as to why we decide to recommend a given movie or show," they add. "*We are not recommending it because it suits our business needs, but because it matches the information we have from you*: your explicit taste preferences and ratings, your viewing history, or even your friends' recommendations [emphasis added]."

Transparency and trust-building are central to how Netflix recommendation aligns user experience and user expectations. Most users will take a chance—and spend time—with recommenders they trust. Consequently, recommendations are experienced less as promotional tools than personal invitations.

By relentlessly remaking itself as a streaming personalization platform, Netflix created, cultivated, and celebrated a brave new kind of customer: the binge viewer. You can't scale binge viewership without simultaneously scaling trusted, personalized recommendation. But when you have rising numbers of binge viewers, you get the essential data and analytic insights for trusted personalized recommendation. Binge viewing seamlessly flows from recommender experience.

"Netflix's brand for TV shows is really about binge viewing," said Netflix CEO and cofounder Reed Hastings back in 2011. "It is to accommodate, to just get hooked and watch episode after episode. It's addictive, it's exciting, it's different."[4] The rise of binge viewing correlates perfectly with the Netflix's own rise in value and impact on Hollywood.

Contemporaneous surveys suggested that over 60 percent of Americans described themselves as "regular" binge viewers. What's more, almost three-quarters of those surveyed said that watching several episodes of the favorite shows in a row made the program experience more enjoyable.

Note the unsubtle Netflix agenda here: personalization isn't limited to suggesting shows customers might enjoy; in Netflix's ontology and teleology, it means recommending shows customers might enjoy binge watching. That's a hugely different expectation, aspiration,, intent and outcome Because they know their customers so well, Netflix's recommenders confidently seek a commitment. Because Netflix's subscribers trust the recommendations, they're often prepared to make one.

Choice Architecture

The recommender experience is essentially a story—a narrative—about choice. Behavioral economics, by contrast, tells stories about choice architectures. That's why all recommendation engines, whether by explicit design or tacit default, are creatures of behavioral economics.

Behavioral economics is the Nobel Prize–winning study of how psychology—thoughts and emotions—influences and affects economic decision-making. Its origins and insights are brilliantly described in both 2002 Nobel laureate Daniel Kahneman's *Thinking, Fast and Slow* and 2017 Nobel laureate Richard Thaler's *Misbehaving*. The essential takeaway: people can be "predictably irrational" about the decisions and choices they make. Seemingly trivial changes in how choices are presented or structured can lead to dramatically different decision outcomes. A gentle nudge at the right moment can change everything.

All recommendation engines, whether by explicit design or tacit default, are creatures of behavioral economics.

That's why choice architectures matter enormously. All recommendation engine designers are inherently choice architects.

Choice architecture is best defined as "the design of choices with the goal of influencing decisions." This includes the user interfaces, associated text, imagery, and sound—as well as structure—of choice presentation. Unsurprisingly, the definition works well as a functional description of a Netflix, Amazon, or booking.com recommender system. While a choice may not always be a recommendation, a recommendation always presents a choice. Ambitious choice architects have no shortage of powerful, provocative, and practical psychological heuristics, such as framing and anchoring effects, to explore and exploit for their designs. Pun intended, this research and its findings have proven remarkably persuasive.

These heuristics are the metaphorical building blocks for choice architects. What kind of recommendations do they want their users to experience? Thaler and his Harvard Law collaborator Cass Sunstein, for example, champion recommendations that "nudge." Nudges are recommendations rhetorically engineered to live somewhere between a hint and a shove.

A "nudge," the two explained in their eponymous 2008 bestseller, "is any aspect of the choice architecture that alters people's behavior in a predictable way without forbidding any options or significantly changing their economic incentives. To count as a mere nudge, the intervention must be easy and cheap to avoid."[5]

Successful nudges are typically inspired by empirical and experimental insight into reliably quirky decision behaviors. A Virginia supermarket, for example, crafted a cheap and easy nudge to encourage healthier eating. Every shopping cart was divided in half with a bright strip of yellow tape. A sign in the cart asked shoppers to put their fresh fruit and veggies in the front half of the cart, and everything else in the back. When shoppers saw the stark imbalance between their fresh and processed food selections, their behavior measurably changed. Produce sales reportedly doubled during this healthy eating promotional nudge.

"Social proof" is another data-enabled, behavioral economic nudge that exerts disproportionate decision impact. Collaborative filtering recommenders—"people like you; . . . people who bought *this* also bought . . ."—are the paradigmatic social proof choice architecture. Similarly, how Airbnb, booking.com, and other hospitality sites display how many people are currently viewing a particular property (for example, *three people are looking at this room*)—and/or highlight how many people have already booked a particular room (for example, *twelve people have booked this room type for your selected dates*) reflects social "nudge-ability." The quasi-rational influence of social validation, "herd behavior," and fear of missing out converts many browsers into impulse buyers. A little nudge can give a recommendation a big boost.

"I think of nudging as like giving people GPS," Thaler has noted. "I get to put into the GPS where I want to go, but I don't have to follow her instructions."[6]

But why not "go with the flow"? To keep its drivers busy and limit defections to Lyft, for example, Uber uses an algorithm analogous to Netflix's binge-facilitating "autoplay" feature. Much as Netflix automatically queues up the next episode of whatever series you're watching, Uber drivers are shown—recommended?—their next possible fare before they're done dropping off their current rider. Nudges prompt; they don't compel. They "feel" helpful.

Other choice architectures rely on "decoy effects" to make reasonable recommendations seem better than they really are. Also known as "asymmetric dominance," the decoy effect exploits a cognitive bias in how most people evaluate choice. The canonical decoy effect vignette come from an *Economist* online sales promotion. The *Economist* displays three subscription options—a digital subscription for $69.00; a print subscription for $119.00 and a print + digital subscription also for $119.00. Who *wouldn't* choose the dual option? It looks like the best value on offer.

Of course the *Economist*'s choice architects *know* this offer appears irrational—two unequal options being sold for the same price. But they also know their middle option decoy makes that third option look reasonable. Removing the decoy clarifies the reality. An option is asymmetrically dominated when it is inferior in all aspects to one option but inferior in some respects and superior in other respects to the other option. If an asymmetrically dominated option is present in a consideration set, consumers will be far likelier to choose the largest (most expensive) option. Just as with their real-world hunting counterparts, digital decoys demonstrably work.

The predictable irrationality of asymmetric dominance applies to decoy recommendations, as well. Choice presentation matters. Does asymmetric dominance truly qualify as a nudge? Technically, yes, albeit a rather slippery one. Nudges inherently raise ethical concerns. Recommendations rooted in human faults and foibles may feel—and be—more manipulative than helpful. Nudges reflect the motives of their makers. "There's a sea of moral dilemmas," agrees Daniel Read at Warwick Business School's Behavioral Science Group.[7]

Nudge champions and defenders assert that there are no "dark nudges," only "dark nudgers." Exploitive choice architects are tantamount to slumlords. Richard Thaler—who inscribes his books with "Nudge for good"—characterizes these less benign, more exploitive nudges as "sludge." He articulates three principles of nudging:

- Nudges should be transparent and not misleading.

- It should be easy to opt out of nudges.

- Nudges should improve the welfare of those being nudged.

N.B.: These principles would be valid and compelling even if one substituted "recommendation" for "nudge." That's no accident. Nudges implicitly recommend and recommendations explicitly nudge. Thaler, Sunstein, and Kahneman all insist transparency and trustworthiness are key.

Defining principled choice architecture becomes even more important for the future of recommender experience. As personalization platform technologies improve, data-driven algorithms won't just better customize recommendations, they'll customize the nudges as well. Recommendation engines will learn, for example, what social proof and decoy nudges best bolster individual user recommendations. Recommenders will computationally mix and match nudges to maximize their user impact. With increased capabilities, what choice architecture principles—if any—appropriately guide effective ethical design? Helping users better clarify their own best interests could be both a design principle and an ethical one.

Indeed, academic research strongly suggests recommendation engines don't reduce the time people take to complete a decision making process. On the contrary, recommenders stimulate users to explore more alternatives before finalizing their choice. "Users' perception of the elapsed time is not related to the larger number of explored choices," researchers concluded.[8] Good recommenders often prompt greater self-awareness, not swifter agreement.

For "choosability" researcher Anthony Jameson, this means going beyond the nudge. A "good choice process" is his recommender experience design challenge. He notes people typically want to be able to justify—not just rationalize—a choice to themselves and others. As Netflix demonstrates, recommendation engines can be quite good at generating reasonable explanations. Perhaps recommenders could—or should—compute different types of explanation: justifications that make the user feel good and/or be used to justify the decision to someone else. Ethical choice architects should design with post-recommendation after-effects in mind. Soothe the possible discomfort of post-binge-viewing hangovers, for example.

Similarly, behavioral science research and personal experience indicate most people dislike making hard trade-offs in decision: speed vs. comfort; time vs. money; location vs. convenience. Computationally, though, recommenders could normatively personalize and prioritize those trade-offs for users. Would recommendation engines shielding users from stressful trade-offs—while still facilitating good choices—faithfully adhere to articulated Thalerian principles?

In that spirit, recommendation prophylactics seem a healthy, hygenic choice. "Personal" recommendation engines or "good choice process filters" could scrape out the sludges and screen out dodgy asymmetric dominance decoys and defaults. That suggests a world where users, in effect, choose choice architectures for their own good choice process and recommenders for good recommender experiences. To wit, personal meta-recommendation engines that might help people better manage their Netflix, Uber, Amazon, and *Economist* interactions.

Much as war is too important to be left to the generals, Jameson's overarching thesis suggests recommendation is too important to be entrusted to the choice architects and recommendation engine designers. How people actually experience recommendations—for better and worse—truly is as important as the recommendations themselves.

In this respect, recommenders are rhetorical constructs: users frequently experience recommender systems more as "persuasion platforms" than prioritized presentations of "better" choices. Aristotle's influence on both the rhetoric of design and the design of rhetoric has been profound.

Where recommender systems emerged from information filtering research and nudges from behavioral economics, persuasive technologies came out of Silicon Valley's Stanford in the 1990s. B. J. Fogg, an experimental psychology grad student, called his nascent field "captology" for "computers as persuasive technology" and launched the school's Persuasive Tech Lab. "As I see it," Fogg wrote, "persuasive technology is fundamentally about learning to automate behavior change."[9]

Fogg defined persuasion as "a non-coercive attempt to change attitudes or behaviors": captology emphasizes voluntary change that rules out coercion or manipulative tricks. Where Sunstein and Thaler designed nudges, Fogg proposed "hot triggers"—recommendations that link immediately to desired outcomes. (Amazon's "one-click ordering," patented in 1999, is an example.) "Put hot triggers in the path of motivated people" was his choice architecture design mantra.

His Stanford stature, location, and laser-like focus on persuasion proved both persuasive and influential. As reported by the *Pacific Standard*, Fogg's "famous 2007 'Facebook class,' which pushed students to design and launch Facebook apps at break-neck speed, launched the careers of many of its 75 students, and made some of them big money before they'd even completed the course."[10]

As Fogg recalls, "When our 10-week academic term was done, the applications built by students had engaged over 16 million users on Facebook. A few weeks after class ended, I once again added up the impact. The 16 million total had grown to 24 million user installations." One graduate fundamentally rethought his class app and subsequently cofounded Instagram, the phenomenal photo-oriented social sharing platform acquired by Facebook in 2012 for a billion dollars. Instagram has since become a dominant social media platform worldwide.

The essential Aristotelian takeaway: "persuasion" was no longer seen or treated as product feature, service attribute, or digital marketing gimmick, but as central organizing design principle. That is, the user experience was built around persuasion and persuasiveness. Recommendation wasn't an add-on; it was baked in.

This sensibility irresistibly leads to choice architectures and innovation cultures where everything is a recommendation. Every pixel and touch-point is seen and stroked as an opportunity to persuade; every personalization experience becomes a persuasion experience. Nudges, triggers, persuasion, and choices blur.

Recommending Visualization/ Visualizing Recommendation

Netflix, again, makes a brilliant case study of how data-driven blends of choice, triggers, and nudges create personalized persuasive user experience [although TikTok's mix of UX and RX makes its video offer world-class]. The company knew that artwork—the pictures and imagery—of its video content had a disproportionate impact on user choice. Originally, the company used generic title images provided by studio partners, typically scaled-down thumbnail versions of DVD cover art. Their quality and appeal was mixed, however, so Netflix ran experiments to determine which images were most likely to quickly catch people's limited attention. Engagement metrics include click-through rate, aggregate play duration, and what percentage of views had short durations.

"We conducted some consumer research studies that indicated artwork was not only the biggest influencer to a member's decision to watch content, but it also constituted over 82% of their focus while browsing Netflix," said Nick Nelson, Netflix's Global Manager of Creative Services, back in 2016. "We also saw that users spent an average of 1.8 seconds considering each title they were presented with while on Netflix."[11]

In the past, Netflix sought to optimize cover art by determining the most enticing image for the biggest number of users. The company ran thousands of experiments to find the best possible artwork for a title, say, *Stranger Things*, that would win the most plays from the largest fraction of members. Upon reflection, researchers realized this approach optimized the wrong variable.

"Given the enormous diversity in taste and preferences," wrote one researcher, "wouldn't it be better if we could find the best artwork for *each* of our members to highlight the aspects of a title that are specifically relevant to *them*?"[12]

This "artwork personalization" project, said Netflix, was "the first instance of personalizing not just what we recommend but also how we recommend to our members." Its research agenda—drawing extensively upon fundamentally re-envisioned the recommendation experience around the thumbnails: identify imagery that let members find stories they wanted to watch faster; ensure that members measurably increased engagement and watched more in aggregate; and make sure titles weren't misrepresented as multiple images were evaluated. (That is, avoid misleading images that would be counterproductive.)

"Each section of the screen is a precious piece of real estate," asserted Netflix's Basilico, "so we want to ensure the member understands why this particular title is being shown to them, especially for a new original series or movies that we create."

"To do this," he explained, "we select the imagery based on the person's viewing habits. For example, if we take a romantic comedy, maybe you've watched a lot of comedies so we show you a funny image from that movie. If another person watches more romances we may use an image of a couple from the movie on a date. It's about helping a member understand what's in it for them by presenting the most appropriate imagery we have."[13]

"Let's say *Good Will Hunting* is recommended," Basilico colleague and collaborator Tony Jebara added. "If we know someone really loves comedies because they've watched

Zoolander and *Arrested Development*, the title image might have a picture of Robin Williams."[14] Is this bespoke imagery more nudge or hot trigger?

Netflix's extensive—and ongoing—personalization experimentation discovered that artwork featuring recognizable—or polarizing—characters from the title tend to do well. In practice, Netflix identified how the art might be edited or curated to maximize individual impact. Pithily put, Netflix's research found successful thumbnails should show close-ups of emotionally expressive faces, show villains instead of heroes, and avoid showing more than three characters. Reengineering the artwork's choice architecture delivered the desired improvements in customer engagement metrics. Binge-worthiness increased.

This artful measure of success here couldn't be further from the million-dollar metric that determined Netflix Prize. Back in 2009, competing researchers sweated the statistical details of RMSE—reduce mean square error—to improve recommender ratings prediction accuracy. Fewer than five years later, Netflix explicitly acknowledged that user experience mattered most. At the 2014 RecSys academic conference, the company's then-head of algorithms put up a PowerPoint slide with a circle/slash through RMSE as a "key performance indicator."

More importantly, Netflix's visualization recommendation innovation literally—and figuratively—illustrates a larger personalization point: the medium matters. On laptops or tablets, personalized thumbnails can effectively convey affect; on smaller mobile phone screens, not so much. Physical constraints and form factors "procrustes" the choice architect's abilities to imbue that "precious piece of real estate" with greater recommendation value. At some point, 'space available' forces thumbnails to shrink too much and/or be nudged off-screen. There's simply not enough room.

As Mediaan, a Dutch data science company, observed in a nifty "teardown" of Netflix presentation algorithms, the company takes extraordinary pains to personalize every pixel of screen real estate it can. Every row has a reason and every reason has its row:

The personalized video ranker This algorithm orders the entire catalogue of videos for each member profile in a personalized way. It is responsible for the "genre rows" (that is, rows like Movies with a Strong Female Lead).

Top N video ranker This algorithm produces the recommendations for the Top Picks row, it finds the best few personalized recommendations in the entire catalogue for each member.

Trending Now This algorithm detects the "short-term trends" for the Trending Now row. There are two types of trends: yearly trends (such as Halloween or Christmas) and one-off events that spike interest in certain categories or movies (a hurricane spiking an interest in natural disaster documentaries).

Continue Watching This algorithm ranks the videos in the Continue Watching row based on how likely you are to resume watching a certain video.

Video-Video similarity This algorithm deals with the Because You Watched . . . rows. This is a two-part process; with the first part being the generation of a list of similar videos for each video in the catalogue (which is unpersonalized), and the second a personalized ranking of each video within the row.

Page Generation: Row Selection and Ranking This algorithm is responsible for generating the entire page (that is, which rows to show where in the page based on relevance and diversity).[15]

An "inverse Bauhaus" principle applies: instead of "form follows function," "form factors influence influence." The physical limitations and constraints of devices define and determine recommendation delivery. This is unambiguously clear for image-oriented recommenders, be they book covers or hotel lobbies. Users looking for dates and hook-ups, however, want unambiguously clear and authentic pictures of who is available. People need to see their choices onscreen.

Instead of "form follows function," "form factors influence influence."

Comparable design issues apply throughout the sensorium: how should musical nudges and choices be presented? What's the best UX for skipping a song or abandoning a playlist? A recommender-oriented automobile GPS system might acoustically recommend in real-time the fastest route, the route with the best views, or the route with closest access to a public bathroom.

This poses particular UX challenges for voice-driven services such as Alexa and Siri. "Play a happy song" is a different request than "Recommend a happy song to cheer me up." Do Alexa or Siri make choices clearer when they make recommendations instead of simply giving answers? How might Siri or Alexa acoustically nudge their users? At some point, a Siri or Alexa might quietly whisper suggestions via Bluetooth or Apple Airpods.

Amazon has explicitly declared its strategic intention to rise to this multimodal recommendation challenge. In 2019, Amazon executive David Limp described a nascent but fast-growing "Alexa economy" that would extend the company's recommendation expertise throughout its digital ecosystems.

For example, Amazon launched proactive Alexa recommendations in 2018 with Alexa Hunches. Initially, Hunches served primarily as a reminder and/or notification service—telling users if they had left a light on that they typically turned off at night. Hunches was upgraded to recommend automating certain tasks based on

Alexa-monitored user habits or behaviors. These Alexa'd automations are called "routines."

So if you set an alarm or check your commute on a regular basis, Alexa might suggest you add it to a routine. Or if you put on the news after checking the commute, Alexa might ask if you want to create a routine. Building recommendations around alarms, commutes, and news is logical, Amazon researchers told VentureBeat, because these use cases typify what most people do most often with digital assistants.

Routines are particularly important to Alexa's "recommendation roadmap" because a key learning from "smart speaker" growth was that adoption rates are higher for users who regularly incorporate features or voice apps into their daily lives. That regularity creates predictability and that predictability invites recommendation.

"As an AI scientist, it's one of the hardest problems to get right," Alexa AI chief scientist Rohit Prasad told VentureBeat in 2019. "This is where you're taking the friction away and dynamically suggesting routines and completing through voice, but again, I want to set the expectations here: [It's an] exciting feature, but this will be the one where we keep learning a lot."

Prasad added that Amazon's wearable devices would leverage their situational and contextual awareness to make recommendations outside the home. Echo Buds, Amazon's Alexa-enriched earbuds, could "chirp in your ear" about [Amazon-owned] Whole Foods' inventory while users shop in-store. "If people actually use this feature," he observed, "proactive voice recommendations like sales or coupon offers might not be too far behind."

Those voice recommendations may be "celebrity-flavored" to make them even more intriguing. Amazon announced that celebrity voices—such as Samuel L. Jackson's—may become an Alexa option for reading weather updates or telling you what's on your calendar. Having users customize what personalities they want proactively recommending what to do next, for example, could powerfully influence nudgeworthiness. Celebrity voices selected to proactively encourage fitness activities, for example, would likely sound different from those recommending dining and/or music experiences.

Amazon's increasing reliance on voice recommendation-and-response understandably prompts serious investment and experimentation in "frustration detection." Real-time recognition of frustration and other tonalities is part of a larger research initiative to understand human emotions. To improve Alexa and its hunches, Amazon already uses LSTM—Long Short Term Memory—neural networks to monitor tone of voice and actual words. These help determine if users are 'happy' with Alexa query results and recommendations.

"The most powerful and transformative part of Alexa's push into people's lives is recommendations—the ability to predict your needs and remove all friction to fulfill them," VentureBeat concluded.

As devices relentlessly evolve—as visual, acoustic, haptic, and even olfactory capabilities are digitally integrated into everyday moments— algorithmic opportunities for more personalized recommendation combinatorially expand. Multimodal, multisensory recommenders become brave new [virtual] worlds for choice architects.

"Recommendations and personalization live in the sea of data we all create as we move through the world, including what we find, what we discover, and what we love," observed Greg Linden, who successfully pioneered Amazon's earliest recommendation engines, in his thoughtful 2017 retrospective. "The field remains wide open. An experience for every customer . . . offering surprise and delight . . . is a vision none have fully realized. Much opportunity remains to add intelligence and personalization to every part of every system, creating experiences that seem like a friend that knows you, what you like, and what others like, and understands what options are out there for you."[16]

RECOMMENDATION INNOVATORS

While individually unique, Spotify, ByteDance, and Stitch Fix illuminate common but transcendent recommender themes. They're organizationally and culturally committed to cultivating data as an asset; their platforms are engineered around continuous learning; and these firms are dedicated to delivering novelty, serendipity, and surprise for their users. Leadership doesn't just "get" it; they demand it.

First-generation internet start-ups were "born digital"; these companies were "born recommending." Their purpose, business models, and growth trajectories cannot be divorced from the quality of recommendation experiences they provide. These platforms are always—always!—experimenting for and with their users; discovering new ways to enable discovery is a strategic aspiration. These firms passionately embrace machine learning and artificial intelligence as essential to that strategy. Technically, they grasp—and effectively ensemble—any algorithms they can to better personalize and customize. Commercially, they strive—and struggle—to strike a profitable balance between business imperatives and their users' best interests. How users recognize and respond to recommendations determines success. Irresistibility is the coin of their realm.

Certainly, Facebook, Alibaba, LinkedIn, Pinterest, Netflix, match.com, or Amazon would make comparably excellent case studies. But these three—each with profoundly different founders, cultures, and capabilities—collectively capture what recommendation engines must do to engage users. Importantly, and provocatively, these companies embrace both social and individual in their purpose and aspiration. That is, they expect recommendation to empower and influence individuals even as they aspire to measurably shape the values and aesthetics of their broader user communities: music for Spotify, video content for ByteDance, and fashion for Stitch Fix. In global markets, recommendation is as much about pop culture as commercial profit.

Spotify

Enjoying roughly two hundred million users in 2018, Spotify is the world's largest independent music streaming platform. The Swedish innovator has successfully made discovery, recommendation, and personalization pillars of its growth and user experience strategies. Discover Weekly is the company's premier recommender system.

Each Monday, over one hundred million customers receive a customized mixtape of thirty songs they'd likely never heard before but were probabilistically likely to love. Spotify's Discover Weekly service makes an incisive case study in how rethinking recommendation and ensembling algorithms profoundly changes people's paths to novelty. Quietly launched in 2015 as a bootleg project, Discover quickly became a hit, reaching a billion recommended tracks streamed in its first ten weeks.

The critical insight: playlists, not individual songs and music, should be the organizing principle and platform for discovery and recommendation. "When I joined [Spotify] in June 2013, I was on a team that was building the initial discovery product . . . it was 'content' in an almost Pinterest-style layout," recalled Spotify engineer Edward Newett, a Discover cocreator. "At some point, a colleague and I decided that it would be a lot easier if we had it as a playlist."[1]

"Playlists are the common currency on Spotify," said Matthew Ogle, who came to the Discovery service from This Is My Jam, a failed startup that asked users to pick favorite songs one at a time. "More users knew how to use them and create them than any other feature."[2]

With roughly seventy-five million Spotify users at the time, Newett and colleagues began mapping the technical challenges of creating seventy-five million individual playlists. Early prototypes merged machine learning tools with recommender algorithms. There was plenty of music to work with: over thirty million songs and one billion-plus user-generated playlists.[3]

Similarity remained Spotify's secret sauce. The system examined their users' behaviors and identified the critical commonalities between songs and artists, crawling user activity logs, user playlists, music news and reviews from around the web, and raw audio files—spectragrams—for capturing such things as tempo, key, and loudness. The service then screened out music users have heard before and sent out bespoke playlists. Songs can come from professionally curated playlists or from regular users. If Spotify notices that three songs tend to appear on playlists together, and you have only two of those songs on a playlist, it will suggest the third to you.

Spotify comprehensively individualizes user profiles of musical preferences, grouped into clusters of artists and micro-genres—not just "rock" and "rap" but fine-tuned distinctions like "synthpop" and "southern soul." Echo Nest, a MIT Media Lab "music intelligence" spin-off acquired in 2014, furnished software to identify emerging genres by crawling music sites and analyzing artist descriptions.

Discover Weekly's truly essential ingredient is the taste and sensibilities of other people.

That said, Discover Weekly's truly essential ingredient is the taste and sensibilities of other people. Different algorithms address different features, attributes, and elements in the computational pursuit of predictably pleasant surprise. More technically, the algorithmic ensemble consists of

1. Collaborative filtering algorithm—nearest neighbor—that finds users who are similar to each other based on listening history—songs they've all listened to—and recommends songs that "Spotify music lovers like you" liked.

2. Natural language processing techniques—such as Word2Vec—mathematically represent implicit relationships, associations, and co-occurrences between words. They structure and analyze playlists as if they were paragraphs or big blocks of text and then treat each song or song title in the playlist as if it were an individual word. As the name implies, Word2Vec converts words into vectors whose distances—similarities—can be measured. If two words (or song titles) appear frequently in the same context, for example on the same playlist, they should be representable by two nearby vectors.

Given a word, the model tries to contextually predict surrounding word representations. In recommender terms that means word2vec will infer high similarities between words that can be used in the same manner. That makes the semantics of music playlists comparable.

Another application creates a "music taste" vector for a user by averaging together the vectors for songs they like to listen to. This taste vector can become the query for a search to find songs that are similar to the user's taste vector. One vector represents meaning; the other captures preference.

3. Outlier/anomaly detection. By definition, outliers are extreme values that dramatically deviate from other observations. They may indicate a variability in a measurement, experimental errors, or a novelty. Commonly used in financial security—it's how banks and credit card companies detect fraudulent charges—outlier detection can have special relevance to recommenders.

Outlier detection determines if a particular instance—that is, listening to a song—is part of a normal behavior pattern or not. Is that isolated "off-the-wall" song a one-off or a mistake? Or might it signal a potential recommendation opportunity? That's a playlist experiment waiting to happen.

4. Deep learning/convolutional neural network (CNN) that process the acoustics—the spectrograms—of songs to identify underlying similarities in acoustic patterns.

CNNs are known to give great results in the area of computer vision—picking out faces in photographs, for example; this pattern-matching success was mapped to sound. The neural network learns the features of songs that make them more—or less—likely to belong to one genre or another.

These networks learn to recognize different features of the input; when these simpler features are stacked on top of one another, more complex features are learned. These neural networks consist of a convolutional layer followed by pooling layer.

To oversimplify, a convolution layer is where a filter initially converts data into features; a pooling layer—which is typically sandwiched between convolution layers—simplifies the information in the output from the convolutional layer. In other words, successive layers computationally refine and enhance the features used to identify the desired/targeted patterns. For example, when we see a photo of a cat, we can appropriately classify it if the photo has identifiable features such as whiskers or four legs. Similarly, neural nets perform image classification by looking for low level features such as edges and curves, and then abstracting them to concepts through a series of convolutional layers. For Spotify, the CNNs are effectively trained to identify the acoustic features that matter most.

5. Finally—and most obviously—how much the user liked or listened to the songs in the previous week's Discover Weekly playlist.

"On one side, every week we're modelling the relationship of everything we know about Spotify through our users' playlist data," Ogle summarizes. "On the other, we're trying to model the behavior of every single user on Spotify—their tastes, based primarily on their listening habits, what features they use on Spotify and also what artists they follow. So we take these two things and every Monday we recommend what we think you would like, but might not have heard about."[4]

When testing Discover Weekly in early 2015, the bootstrap team quietly pushed its prototype to the Spotify accounts of all company employees. "Everyone freaked out in a good way," Newett said, saying things like "'It's as if my secret music twin put it together.'"[5]

To make visually clear that this playlist was personalized, they decided to illustrate each with an image of the user, pulled from Facebook. And they determined that they would refresh the lists weekly, on Monday mornings.

"We were feeling good, but we hadn't tested it on [typical] users, so we then rolled it out to one percent of the user base," he recalls. Again, enthusiasm reigned.

Then came the mid-2015 rollout to the rest of Spotify's one hundred million active users around the world.

"We had to refresh 100 million playlists every Sunday night, with about a terabyte of new data." The ensemble delivered.

"This wasn't a big company initiative," Newett recalls, "just a team of passionate engineers who went about solving a problem we saw with the technology we had."

ByteDance

Launched in 2012, Beijing ByteDance quickly leveraged an "everything is a recommendation" network architecture to become one of China's fastest-growing innovators. Unintimidated by the scale and success of platform giants Alibaba, Baidu, and Tencent (of WeChat superapp fame), the company's commitment to customized content fiercely disrupted China's digital elite. ByteDance beat its rivals by creatively marrying artificial intelligence to continuous personalization.

"ByteDance has a variety of products that are content platforms powered by machine learning," says Liu Zhen, ByteDance senior vice president for corporate development. "It's really a discovery platform that provides personalized content to the audience."[6]

With over 120 million daily active users in China (at this writing), Toutiao, the pioneering news aggregation app is the company's best-known service. Douyin—called

TikTok outside China—is an addictive short-form mobile video platform that's successfully gone global. During the first half of 2018, TikTok became the world's most down-loaded app on Apple's App Store with over 104 million downloads. The interrelated evolution of these apps make clear China's aspirations as a global exporter of recommendation innovation.

Strategically and opportunistically, ByteDance aligns ongoing algorithmic innovation with simple, fast, and intimate user experience design. Maximizing measurable engagement is key. The average Toutiao (that's "headlines" in Chinese) user spends more than *seventy-five minutes* each day on the app—more than what average users spend on Facebook and double their time on Snapchat. According to one analyst, that makes Toutiao "one of the most addictive and sticky apps in China."

Over half that time is spent watching short-form videos (which helps explains ByteDance's TikTok foray). Coupled with over ten billion video views per day, this multimedia format has made Toutiao a virtual YouTube of China. Engagement intensity correlates with growth. Within four months of launch, the service reached a million active users. Daily active user numbers accelerated from thirty million in 2015 to seventy-eight million in 2016 to over 120 million in 2018. Toutiao's motto: "The only true headlines are the things you care about."[7]

"We strive to become the information platform that knows each individual best," Toutiao vice president Tina Zhao told the *Financial Times*. "No two users' feed lists are alike."[8]

This growth, engagement, and personalization combination redefined China's advertising for the news aggregation market: in 2017, Toutiao's ad revenues topped $2.5 billion, almost tripling 2016 levels. By 2018's end, the parent company was described as the most valuable start-up in the world.

From its start, Toutiao tracked information about each user—taps, swipes, time spent per article, and location—to power the recommendation engine. No explicit user input is required; unlike with Facebook or We-Chat, recommendations aren't derived from social graphs. Toutiao's data scientists use natural language processing and computer vision capabilities to extract hundreds of entities and keywords as features from each piece of content. More importantly, says ByteDance research vice president Wei-Ying Ma, "We extract hundreds of high dimensional features from users and model user interest based on the data. Users' profiles are updated in real-time after each action."[9]

The data volumes beggar belief: in 2017, Toutiao possessed approximately 190 terabytes of user profile data—the amount of data the Hubble Space Telescope would produce over twenty years. The service processed nearly eight petabytes a day—the rough equivalent of fifty-five billion Facebook photos. There's a growing wealth of data to learn from and the company is totally committed to "the real AI"—algorithmic innovation.

So when users first open—"cold start"—the app, Toutiao pops up preliminary recommendations based on their mobile device's operating system, location, and other factors. Their screens immediately stream posts ranging from traditional news reports to "cute" and popular viral videos. The more clicks and swipes, the more Toutiao learns and refines each feed. The result is personalized, extensive, and demonstrably relevant content. The focus is overwhelmingly on user appeal.

What are the one hundred articles the platform can recommend to each user that are most likely to result in continued engagement? is the essential question. Toutiao's algorithms answers come from learning the digital details of user behaviors; their viewing habits and physical locations. User profile, content, and context interactions generate the training data that rank the most appealingly relevant recommendations. Toutiao optimizes recommendations around two objective engagement metrics criteria: the percent of articles/videos a user views (clicks) and the proportion of content they actually finish (time spent.) According to the company, its algorithms successfully learn user interests in less than a day, given the 80 percent click-though/completion rates.

But this machine-learning goes well beyond the current query, said ByteDance's Ma. Chinese expats using the app in America, he noted, dramatically expand Toutiao's personalization options and opportunities. When Chinese New Year comes around, these users often travel from the United States to China. Their new locations contextually change news recommendation during the holiday season. But once the user returns to America, however, the software has effectively remembered—"learned"—that the Chinese New Year location was likely significant; probably their hometown. So relevant news stories that popped up and were read in their likely hometowns, says Ma, will likely continue to occasionally show up in their American feeds.[10]

This contextual awareness also extends to time of day, he notes. Toutiao's "delivery" algorithms track and learn when users are "busy." The app will throttle the flow of streamed stories and videos and patiently wait until users are most likely to be free. Again, user behaviors and actions are updated in real-time. Toutiao's algorithms are always in training; they're always learning to personalize and customize.

Like Netflix, Toutiao understands machine learning algorithms can create content, not just recommend it. At the 2016 Olympics, for example, Xiaomingbot—a ByteDance machine learning "bot"—wrote sports reports for certain events after analyzing real-time images and

Machine learning algorithms can create content, not just recommend it.

integrating sports vocabulary. Written and posted barely three seconds after the event finished, these "bot-icles" nevertheless enjoyed read rates (# of reads and # of impressions) comparable to those of human journalists.[11]

Technically, this required Toutiao's data scientists to combine natural language processing capabilities with contextual image recognition. Meaningful content creation required "story templates" on which to pour structured narratives; ranking algorithms to select relevant sentences from live text commentary; and image-text matching algorithms to tie them all together. To ensure greater sophistication, convolutional neural networks analyzed candidate imagery. By training on historical data, that is, past juxtapositions of stories and images, the model learned to pick appealing images for the story.

Sequence-to-sequence deep learning algorithms (also known as Seq2Seq) that train models to convert sequences from one domain (e.g., Chinese Sentences) to sequences in another (e.g., translated into English) summarized existing stories into daily highlights and suggested more potentially "viral" titles. Toutiao's system also employed recurrent neural networks (RNNs)—which do for text when CNNs do for images—to compute vector representations for sentences. Those semantic sentence vectors, in turn, are fed to a ranking model that effectively extracts and edits them into readable summaries.

ByteDance continues to work on smarter, personalized sports coverage, according to the company's Ma. The "one-feed-fits-all" approach described above is being superseded by customized viewing experiences. When data indicate interest in, for example, specific players, the recommenders compile and customize the content accordingly. So Xiaomingbot 2.0 would highlight those players, with bespoke, automated commentary and captions.

As in America, machine-driven recommendation innovation inevitably leads to content innovation. "We adjust our strategy every week. It's a constant experiment," said Ma, who adds that machine learning can also help predict the content "virality"—whether video or text—throughout Toutiao's media ecosystem.[12]

This recommendation/innovation sensibility is embedded deep into the Douyin/TikTok app experience. First launched in China in the fall of 2017, the short-form mobile video platform was developed in two hundred days, reached one hundred million users in its first year, and quickly enjoyed over a billion views a day.

Like Toutiao, Douyin/TikTok immediately opens to recommendations. Unlike Instagram, where content comes from people you follow, TikTok displays video content users are unlikely to have seen before. Discovering appealingly novel content and creators is the goal. "The important design choice here is to launch the user into the 'For You' tab," observes one technically sophisticated reviewer. "Swipe to the next video—forever—Swipe up or down, to move to the next/previous video. Or Swipe in from right to pull up a users profile. No gesture goes unrewarded. 'Sorry! This action is not supported'—is something you'd never see on this app. There is no play or pause button on the app. Instead, you can keep swiping up for the next video in an infinite line of 15-second clips."

This relentless blend of personalized recommendation and ease-of-use has made Douyin/TikTok almost as addictive as its Toutiao predecessor. According to the KPCB 2018 Internet Trends Report, TikTok has a remarkable daily average user/monthly average user "stickiness" ratio of 57 percent.[13] Users reportedly spend about fifty-two minutes on the app daily. That's a lot of fifteen-second videos. "Douyin's user stickiness reflects its 'Bytedance gene,'" says China Channel mobile analyst Samin Sha.[14]

Unlike Toutiao, though, Douyin/TikTok was born to be both creative and social. Video creation and sharing, not just consumption, is encouraged. Douyin/TikTok turns mobile phones into mobile production studios. Not unlike Instagram and Snap, filters, special effects, and easy editing tools abound. The same artificial intelligence, image capture, and machine learning algorithms supercharging Toutiao have been repurposed for user-generated content creation. Special effects include shaking and shivering with hip-hop and electronic music, hair dying, 3D stickers, and props. Indeed, users posting to TikTok must select music. Ordinarily, music is optional for video uploads; on TikTok, it's mandatory.

Every sound that gets uploaded to the network for the first time gets designated an "original sound" and tagged to the Douyin/TikTok ID of the user who first uploaded it. So users watching videos with a sound they like can tap the Original Sound icon to see every video made with that sound. Another click and the user can see who first uploaded the sound. That leads directly to other discoveries and recommendations. Douyin/TikTok's machine learning algorithms get better trained on what blends of sight and sound stimulates virality. That learning makes advertisers happy, too.

In fact, the stickiness of its success has become a major regulatory ByteDance challenge. China's authoritarian government criticized the company for promoting addictive behaviors and frivolous content. In response to critics, ByteDance has added alarms that ring every half-hour of use; users can block the app after two hours of use. Unlocking the app requires a password. ByteDance's addictive brilliance in recommendation and personalization has invited aggressive regulatory scrutiny even as it has facilitated global growth.[15] Might ByteDance use machine learning for recommendations on how to better cope with regulators?

Stitch Fix

When Stitch Fix went public in 2017 as a NASDAQ unicorn, *Elle* magazine observed, its thirty-four-year-old CEO Katrina Lake was smartly dressed in a burgundy shift and heels. Those sartorial details say less about sexist editorial standards than the algorithmic style and substance of Stitch Fix's success. Ms. Lake is a customer, not just a cofounder, of her company.

Launched in 2011 as an online subscription and personalized styling service for women, Stitch Fix grew to billion-dollar status by blending—and scaling—data science with fashion sense. In 2016, the company sold $730 million worth of clothing; $977 million in 2017; and over $1.25 billion in 2018.[16] Stitch Fix wants its brand to clothe men and families, too.

Bundling analytics, advice, and nudges into "choice architectures" that inspire both customer loyalty and copycat competition, the company—like its founder—is unabashedly data driven. "Data science isn't woven into our culture; it *is* our culture," says Lake. "We started with it at the heart of the business, rather than adding it to a traditional organizational structure, and built the company's algorithms around our clients and their needs. . . . Data science reports directly to me, and Stitch Fix wouldn't exist without data science. It's that simple."[17]

That commitment informs every link of the Stitch Fix value chain. Unlike a Spotify or ByteDance, Stitch Fix's chosen UX imposes touch-points and tangibility that go well beyond sampling music or videos. Cultivating a personal style requires a different kind of data-driven interaction. The company literally and figuratively asks a lot of customers in its efforts to discover and deliver apparel they're most likely to love, keep, and wear.

Stitch Fix customers must fill out an online or "in-app" questionnaire detailing their personal preferences, sensibilities, sizes, and budget. Style Shuffle, for example, was a 2018 in-app initiative to "game-ify" data elicitation by having shoppers thumbs-up/thumbs-down pop-up imagery of clothing items and accessories. As a result, claims Stitch Fix CTO Cathy Polinksy, "We're seeing clients are keeping more items when they're playing the game and they spend more with us over the course of the year."[18]

Drawing on its inventory of roughly seven hundred brands—and Stitch Fix's own labels—an algorithm generates a bespoke list of recommendations for review by one of the company's 3,500+ stylists on call. That stylist picks out five items—a "fix"—to send on a monthly, bimonthly, or quarterly basis, along with a personal note describing how best to style and/or accessorize the selected items. "Data-enriched" stylists, Stitch Fix insists, bring an essential human touch to the selection process. Stitch Fix charges a $20 styling fee for each fix.

After getting their parcel, customers are asked to review each piece online, keep whatever they want, and return what they don't. The $20 fee is applied to any item customers choose to keep. (As a clever nudge-worthy incentive, keeping the entire fix earns a $25 discount.) These data are all logged on the customer's profile to better tailor future fixes. Stitch Fix assiduously monitors hundreds of personalization, product, and merchandising data points in pursuit of perfecting portfolios.

"On the very first Fix, we start getting feedback as customers can try these clothes on in the privacy of their own home and ask their friends and loved ones about different pieces," says Stitch Fix chief algorithm officer Eric Colson, who had been Netflix vice president of data science and engineering. "This is really rich feedback, since they can evaluate clothes based on how they actually fit and look on their body instead of just making a simple decision about what they see on a page in a magazine. Customers will give us some structured information, such as feedback about the fit, the price, the style . . . as well as unstructured information—we have a text box they can fill out with any additional thoughts, like '*this fits great on the body portion, but it's a little tight on the shoulders.*' This . . . detail is incredibly helpful as our algorithms learn a customer's preferences."[19]

Images, however, elicit more valuable feedback than words, Coulson observes; for example, the company invites customers to create fashion albums from Pinterest that lets them "demonstrate what their style is without having them articulate it." But Style Shuffling and Pinterest pictures don't go far enough to seal the deal. They perform better as data lures than reasons to buy.

"I think one of our most interesting findings is that people just can't seem to judge clothing from images alone," says Colson. "You can recommend something on a web page but the customer might just say, 'No, that's not for me.' But if you send it to them and say, 'Trust me; try it on'—which we can do because we have the data to back it up—often times you'll find that customers are surprised at how much they like it. That bit of knowledge, which we learned in 2012, was astounding to us. So many things that would have been rejected if a person were to just judge from an image—or even walking by it in person—we've found customers really enjoy it if they try it on. We call it 'surprise and delight' and it's been a game-changer for us."

Colson first observed this behavior with his own wife before joining the company: "She got her first 'Fix' and the first item was a scarf—to which she said, 'Ugh, I don't want a scarf.' But then she touched it, tried it on and looked at herself wearing it in the mirror and was surprised [by] how much she liked it. By having it right there in front of her, her opinion about it completely changed."

"Experiential trials," Stitch Fix analytics affirm, create customer connections that window-or-screen shopping and styling advice simply can't. "It pushes customers a little bit out of the comfort zones and can be incredibly effective," Colson declares. "Our customers come to us for the convenience and they stay for this surprise."

In fact, says another Stitch Fix data scientist, the company's customers tend to self-segment into two specific groups: "variations on a theme" clients looking for apparel affirming a signature style and "adventurers" desiring distinctively novel fashion-forward attire. The two segments have wildly different comfort zones and perceptions of variety.

"As we learn more about each client over time, both our algorithms and stylists become more accurate," Meredith Dunn, the company's vice president of styling and client experience told the *Washington Post*. "Our stylists read and digest feedback from clients and our algorithms ingest that data, too."[20]

But for the initial fix, says the chief algorithm officer, "the machine learning happens first." Stitch Fix ensembles all kinds of algorithms—neural nets, collaborative filters, mixed effects models, naive Bayes—"to do a first pass at recommending styles for individual customers."

In theory, Stitch Fix's brand of recommendation should readily lend itself to traditional recommender matrix methods like collaborative filtering. But data-driven realities suggest otherwise. "It's true that matrix factorization is the most intuitive place to start for product recommendations," Colson acknowledges, "but they tend to fall short in the fashion industry because trends change so fast. The stuff we have now will be gone next month, so we have really sparse matrixes. Most clients are getting boxes only on a monthly—or even less frequent—basis. This means we only get feedback on about 5 items per month per client on average. As such, it's better to use other features to inform our recommendations, and our techniques run the gambit from neural networks to mixed effects models."[21]

So for customers who've pinned their styling preferences on Pinterest, for example, recommendations emerge from a two-tier human/machine process. Machines vectorize the pinned images—that is to say, the software converts the clothing imagery into vectors whose features are processed and mapped. Those features/vectors are then compared to Stitch Fix's inventory using convolutional neural network architectures that are particularly adept at image processing. Stitch Fix has used the AlexNet, named for its cocreator Alex Krizhevsky—a breakthrough neural net design that won the global ImageNet Large Scale Visual Recognition Challenge in 2012. This innovative deep learning architecture dramatically improved computer vision accuracy rates. This neural net matches the desired user features with the inventory offerings by determining similarity as measured by Euclidean distance. In other words, very sophisticated mathematics identifies the key apparel features; relatively simple mathematics computes the relevant similarities.

"Once we have this short list of items," says Colson, "we pass this to humans who process ambient information, recognize if the recommended items are too similar to the pinned image, or even judge whether the images are more aspirational than literal and modify suggestions accordingly. Take something like a leopard print dress. Machines are very pedantic: they can distinguish leopard print from cheetah print, but don't have the social sense to know that a woman who likes leopard print would very likely also like cheetah print."

Stitch Fix's division of recommendation labor, says Colson, leverages Nobel laureate Daniel Kahneman's behavioral economics/choice architecture insights articulated in *Thinking, Fast and Slow*. "The machines take the calculations and probabilities," he observes, "the humans take the intuition."

"A good person plus a good algorithm is far superior to the best person or the best algorithm alone," asserts CEO Lake. "We aren't pitting people and data against each other. We need them to work together. We're not training machines to behave like humans, and we're certainly not training humans to behave like machines. And we all need to acknowledge that we're fallible—the stylist, the

A good person plus
a good algorithm is
far superior to the
best person or the best
algorithm alone.

data scientist, me. We're all wrong sometimes—even the algorithm. The important thing is that we keep learning from that."

One emergent learning recapitulates the TikTok creation experience: Stitch Fix's breadth of customer and merchandising data now empowers the company to design, not just recommend, apparel. Hybrid Designs, the company's internal AI-driven design group, ensembles three algorithms to generate clothes: the first selects three "parent pieces" that could be combined or serve as templates for a new piece. The second highlights three attributes that have empirically complemented and/or enhanced the parents' style for Stitch Fix customers—sleeve-lengths or buttons. The third algorithm injects a structured randomness that facilitates design novelty. These combined attribute "Frankenstyles," as they are nicknamed inside the company, do particularly well in certain customer demographics.[22]

Amazon has noticed Stitch Fix's fashion and commercial success. The world's largest retailer, with formidable recommendation and machine learning capabilities of its own, revealed plans that put Stitch Fix, its stylists, and Hybrid Designs directly in its sights through a directly competitive service both in the personal styling and data-driven clothing design phase. Unlike its biblical predecessor, this Goliath learns.

For these companies, recommenders aren't features, functions, add-ons, or a layer of "the stack"; they're central to the enterprise. Innovation, investment, and UX flow from—and into—recommendation. They honor the celebrated Web 2.0 virtuous cycle commandment: the more people use them, the more valuable they become; the more valuable they become, the more people use them. Fashion sense gets keener, viral videos grow more virulent, and catchy tunes get caught. Recommenders are springboards, as well as platforms. Recommendations create new ways to engage with—and add value to—users.

Three key organizing principles underlie their recommender technology success:

1. Recommendation as "prediction" technology. Recommendation is everywhere and always about the future. What should users do—listen, see, wear—*next*? With Spotify playlists and Stitch Fix fixes, those nexts are bundles or portfolios of choices and options. Or, with TikTok, quick, easy, and instant swipes. These predictions are data-driven investments to influence or own a piece of the user's future. How valuable are those investments? Measurably owning a moment, an hour, a day, or a lifetime is the prediction challenge; monetizing those moments is the business challenge. Key issue: can TikTok, Spotify, and Hybrid Design predictions, over

time, fundamentally change people's preferences and tastes?

2. Recommendation as a "discovery process" technology. Predicting what users will like next is a necessary but not sufficient for success. Stitch Fix wins when customers keep—and wear—that blouse they never would have imagined trying on. TikTok's addictiveness increases when users discover feral cat videos are even cuter than tabby cat videos. Spotify's Discover succeeds when subscribers who like music suddenly hear a song—or artist—or genre—they love. Ongoing engagement requires the element—if not a periodic table—of surprise and delight. A recommender's ability to map out and navigate a user's comfort zone is not enough; these recommenders expand and redraw comfort-zone boundaries.

3. Recommendation as a "production methodology." Much as Netflix relied on recommender analytics to inform its *House of Cards* production, these recognize recommendation as a path to product/service innovation. Insights gleaned and gained from recommendation are effectively reverse engineered to algorithmically craft value for users. Stitch Fix's Hybrid Designs breed successful new fashions; Spotify's playlists and raw audio files are mined for earworms and musical patterns with commercial/cultural appeal. ByteDance

increasingly recommends tools and templates that promote video virality for content creators. Virtually every recommender-centric enterprise understands that the feature engineering and latent factors that boost predictive validity can be used to craft more appealing music, more appealing content, and more appealing apparel. In theory and practice, superior recommenders are supremely well positioned to become superior producers.

In this, they recall another algorithmic revolution: the 1970s' rise of the Wall Street quants. Financial deregulation and the diffusion of computing power sparked innovative research in investment mathematics. Data-driven academic researchers—notably in economics and finance—used sophisticated mathematical tools to look for underlying patterns and relationships between financial instruments.

This research would not only expand the boundaries of knowledge around "efficient markets" and evaluating investments, they could be used by investors to make money and better manage risk. Wall Street firms, pension funds, and traders were understandably interested in commercializing this research.

One of the most important research efforts examined the key quantitative challenge of how to value an option. An option is a financial instrument that allows, but does not require, investors to buy an asset, like a stock, at a prearranged price during a fixed span of time. The launch of a new options exchange in Chicago created particular research relevance, opportunity, and urgency to the challenge.

Three economists—Fischer Black, Myron Scholes, and Robert Merton—came up with the breakthrough "options pricing model"—the Black–Scholes–Merton formula. It was introduced in their 1973 paper, "The Pricing of Options and Corporate Liabilities," published in the *Journal of Political Economy,* and it became the trading algorithm that launched a multitrillion-dollar industry. The algorithm empowered people to price, buy, and sell options with greater confidence and efficiency.

But the formula could do more than simply price options, Merton noted; it could also be used to create them. That is, the options pricing model could also be a production methodology. Companies could use the algorithm to precision-engineer options and derivatives of their own. Just like the Black–Scholes–Merton equations, recommender systems not only enhance consumption opportunities, they create the option to produce as well. This double-edged capability has profound implications.

Not incidentally, the Black–Scholes–Merton algorithm proved so transformative that it won the 1997 Nobel Prize in Economics for Merton and Scholes (Black had passed away in 1995 and Nobel Prizes are not posthumously awarded).

Alan Kay, the noted computer scientist at Xerox PARC—where Tapestry and recommender systems were first devised—famously observed that "the best way to predict the future is to invent it." As these transformative examples indicate, that is what recommendation engines will increasingly and innovatively do.

THE RECOMMENDER FUTURE

Self-knowledge. Self-interest. Self-improvement. Self-indulgence. Self-control. The idea and ideal of a "best self." Recommendation engines increasingly shape who people are, what they desire, and who they want to become. That social media prompt suggesting outreach to a forgotten colleague; the remarkably timely article in your feed you would never have otherwise seen; the movie that changes your life; that personal/professional introduction you couldn't have imagined. Algorithmically speaking, recommendation engines generate the personalized options and choices that—with a tap, swipe, or spoken word—turn "next moments" into graspable opportunities. From Alexa to Amazon to Apple to Facebook to Google to LinkedIn to YouTube to TikTok, their pervasive influence can't be escaped. But does their growing power and promise deserve reciprocal embrace?

This final chapter argues that the keys to understanding the recommendation engine future will be found in the future of the self: not a digital or digitized or virtual self but a self radically empowered by the technologies of an algorithmic age. Successful recommenders promote self-knowledge, inform self-interest, nudge self-improvement and invite self-indulgence even as they encourage self-control. They surface and serve "self"-ish needs and latent desires. The recommendations people follow—and ignore—reflect who they think (and feel) and hope they really are and could be.

Like intimate—even intrusive—personal mirrors, recommenders can magnify how people see themselves, their preferences, and their possibilities. They can't help but challenge foundational notions of agency and free will even as they alluringly dangle possibilities and promises of a "best"—or at least "better"—self. Ongoing innovation guarantees recommendation becomes even more powerful and persuasive for defining, designing, and divining one's future self. The future of the self is the future of recommendation.

Recommendation will become fire and fuel for human capital transformation.

The aspirational conceit of recommendation engine pioneers and practitioners is that their precocious progeny will learn to know their users even better than those users know themselves. While that ambition sounds more

Recommendation
will become fire and fuel
for human capital
transformation.

hubristic than idealistic, AI gurus and machine learning mavens expect nothing less. "Discovery should be like talking with a friend who knows you, knows what you like, works with you at every step, and anticipates your needs," said Greg Linden, who led Amazon's earliest recommender work. "A friend who knows you [even] better than you know yourself."[1]

"The algorithms know you better than you know yourself," agrees Xavier Amatriain, a former Netflix and Quora data science innovator.[2] With a touch more modesty, chief technology officer Zhang Chen of China's giant JD ecommerce retailer told the *Economist*, "We will know you as well as you know yourself."[3] This "know you better" trope has hardened into ideological imperative for the digerati. "Supra-self" themes pervade the RecSys academic and research literature. Parity with people isn't good enough; tomorrow's genius recommenders must preemptively personalize even before inchoate desire pops into mind.

But how does such superior self-knowledge manifest? Presumptuousness is an unappealing trait in smart people; it's unclear how charming know-it-all machines might fare. What actually happens when recommendation engines demonstrably know the humans who use them far better than those humans know themselves? Theory and practice both suggest those answers would quickly become clear: users will either be intrigued or delighted by the "know you better than you know yourself" recommendations or that advice would be exposed as not being special at all. That's the test.

More deeply and darkly, however, these questions expose inescapable ethical future self conundrums. Whose interests do those brilliant recommenders truly serve? After all, getting you to happily buy something you had never imagined wanting may not really be to your longer-term benefit. Similarly, how many minutes does it take for those charming video recommendations that pique your curiosity to malignantly mutate into addictive time-wasters? Are recommendation engines more driven to discover and exploit your insecurities than better learn and explore your strengths? Preliminary answers don't flatter the data scientists; unhappy socio-cultural critics see supra-self recommendation temptation as potentially poisonous forbidden fruit. To paraphrase Lord Acton, absolute self-knowledge corrupts absolutely.

For these skeptical historians, philosophers, and futurists, the irresistible rise of uber-recommenders presents a dystopian threat to individual agency and free will. To their minds, recommenders are less slippery slopes than suicidal leaps into digital dependency and addiction. They strongly recommend against recommendation.

"In many cases people will follow the recommendations because they realize from experience the algorithms make better choices," said Israeli historian and futurist Yuval Noah Harari in an interview.

The recommendations may never be perfect, but they don't have to be. They just have to be better, on average, than human beings. That's not impossible because human beings very often make terrible mistakes, even in the most important decisions of their lives. This is not a future scenario. Already we give algorithms authority to decide which movies to see and which books to buy. But the more you trust the algorithm, the more you lose the ability to make decisions yourself. After a couple of years of following the recommendations of Google Maps, you no longer have a gut instinct of where to go. You no longer know your city. So even though theoretically you still have authority, in practice this authority has been shifting to the algorithm.[4]

As that algorithm comes to know you better than you know yourself, Harari contends, "It can predict your choices and decisions. It can manipulate your emotions, and it can sell you anything, whether a product or a politician." Essentially, he declares, you'll be "hacked."

In Harari's unhappy view, recommender algorithms become malware that makes hacking humans not just possible or probable but inevitable. "To hack a human being is to understand what's happening inside you on the level of the body, of the brain, of the mind, so that you can predict what people will do," he states. "You can understand how they feel and you can, of course, once you understand and predict, you can usually also manipulate and control and even replace. And of course it can't be done perfectly and it was possible to do it to some extent also a century ago. But the difference in the level is significant. I would say that the real key is whether somebody can understand you better than you understand yourself."[5]

We've seen that ambition and aspiration before. But which behavioral doors does that real key truly open? Many people have friends, parents, siblings, and spouses who know them well and love them dearly. There's trust, caring, commitment, and connection. And yet, somehow, their best-intentioned, best-informed, and most trustworthy advice gets regularly—if not frequently—ignored. Or they're followed only in the most perfunctory ways. How is that possible? Superior knowledge and trustworthiness are wonderful qualities. But they guarantee neither deference nor obedience. They never have. The thesis that exceptional algorithms will win automatic loyalty and instant obeisance that mere humans can't is unproven and unlikely, not untested.

Empirically, however, the social, psychological, and economic appeal of recommendation and advice comes from optionality, not compulsion. Ignoring good advice can feel even better than following it. The ability—the agency—to just say "No" or "Not yet" or "I need to think it over" is empowering. As for Google Maps–driven gut instincts decline and dependence, Harari's techno-lament recalls Plato's *Phaedrus*'s screed that writing's rise would wound and weaken human memory. *Plus ça change . . .*

More ominously, Harari and his fellow travelers see recommendation engines as algorithmic Trojan Horses that, once inside the neurophysiological gates, subvert and pervert free will. These software seducers offer only the illusion of real choice. "If you believe in the traditional liberal story, you will be tempted simply to dismiss this challenge," he argues. "'No, it will never happen. Nobody will ever manage to hack the human spirit, because there is something there that goes far beyond genes, neurons, and algorithms. Nobody could successfully predict and manipulate my choices, because my choices reflect my free will.' Unfortunately, dismissing the challenge won't make it go away. It will just make you more vulnerable to it. . . . If governments and corporations succeed in hacking the human animal, the easiest people to manipulate will be those who believe in free will."[6]

Instead of dismissing that free will challenge, let's note that, in order to work, those algorithms must also extinguish self-awareness. That is, they must successfully edit or erase people's self-knowledge. In truth, Harari's argument begs the question: a "hacking humans" sensibility presupposes free will as illusory and its most passionate champions as gullible fools. While agency, self-awareness, and free will may be controversial concepts in neurophilosophy, neurobiology, and psychology, Harari and the "recommend-phobes" themselves think not: the science is settled. They freely can't help but believe free will can't exist. How ironic.

But this is a "meat puppet" perspective of the future self: stochastic, probabilistic, and complex systems ultimately yield to dreary determinisms computed by the Googles, Facebooks, Amazons and Alibabas. Not only do they know what's best, their personalized prophecies are self-fulfilling. In recommender-rich, data-driven worlds, digital determinism is destiny. *"I, for one, welcome our new robot overlords."*

Of course, that's why future self/future recommendation engine themes command transcendent concern: the fate of individual expression and human freedom worldwide hinges on their coevolution. So let's be ruthlessly clear about fundamentals: recommendation implies choice. Users—rightly—believe recommendation engines empower them to make and enjoy better choices than they might likely make on their own. But when recommenders know users better than they know themselves, the temptation to be too clever by half becomes unendurable.

Stitch Fix's Eric Colson quotes a product manager from his Netflix days who argued, "If we were really bold, we wouldn't present five or so recommendations. We'd present one and if we [did] that, we should just play a recommendation when the user came online."[7]

If people don't
really have a choice,
how is it really a
recommendation?

Recommendation savants should be careful what they wish for. That brand of "boldness" comes with a semantic catch. If people don't really have a choice, how is it really a recommendation? If users don't have agency—if their options are manipulatively constrained and/or prove inordinately difficult to exercise—is the recommendation authentic? Recommendation engines that offer no real choice—or predominately exploitive ones—are frauds. Even if users ostensibly "like" what they're getting, they literally had no choice. What kind of "selfhood" is that? Like the human pigeons plucked playing three-card monte, they got themselves scammed.

When Harari postulates recommender algorithms effectively hacking your brain or mind and reliably putting you in a trance, we no longer live in the realm of recommendation or advice. On the contrary, we've moved to dark domains that explicitly seek to limit freedom and choice. Free will may or may not be an illusion; denying humans agency is not.

If recommendations imply agency and choice, then addiction or outright dependence accurately describes their absence. Addicts don't choose; dependency devolves into mindless default. Deliberate manipulation, whether by an enterprise or State, inherently deforms and distorts choice. Calling algorithms that deceive and delude "recommenders" thus grossly and dishonestly mischaracterizes them. Their true mission and motivation isn't recommendation or even persuasion; it's obedience, compliance, and control.

That is the literal antithesis of what recommendation engines should ask of their users. Recommenders work with integrity when they measurably empower agency and choice. Recommendation engines designed to disempower users are not worthy of the name. They are liars. At the algorithmic heart of every recommender system worldwide is the essential question made famous by, yes, Rome's master rhetorician and divination skeptic Marcus Tullius Cicero: "Cui bono?" *Who benefits?*

The more recommenders know or anticipate their users, the more important Cicero's question and challenge become. The famously sardonic internet business aphorism applies: "If you are not paying for it, you're not the customer; you're the product." (That said, most successful digital innovators prefer high-quality products. Indeed, why should a Bezos or Brin or Ma want to debase or devalue their most important products?)

States and enterprises dependent on trickery and deceit to achieve desired outcomes, however, have made their intentions clear: self-interest *uber alles*. The ends—political, commercial, cultural—justify their manipulated means. But the recommender future needn't follow Harari's *Black Mirror*–esque scenarios. Recommendation engines can—and do—reflect and respect the self-interest of their users. Principled recommender systems can learn to constructively balance—or correctively bias—conflicting interests. Agency can prove more profitable, productive, and sustainable than exploitation.

Yet *cui bono?* goes only so far: its answers primarily illuminate the methods and motives of the engines' designers. The better and more important question—which haunts Harari's digital dystopians—is "who determines who benefits"? What forces—market or regulatory—ensure that supra-self recommenders minimize malignant hackery and privilege agency over exploitation?

Ultimately, battles over recommender futures—be they dystopic or benign—represent an intensifying clash of political and governance systems philosophies. How could they not? The global stakes are that high. Every policy dispute shaping the future of platforms, privacy, social media, and AI directly affects the ongoing recommendation engine revolution. Asking the traditional "Who shall guard the guardians?" question misses what makes recommenders and recommendation so special. Protectors and champions of agency and empowerment in an era of ever-smarter machines must also answer, "Who shall 'choice architect' the choice architects?" Recommendation, suggestion, and advice cannot be meaningfully divorced from how they are framed. Perhaps "regulatory recommenders" will suggest policy and presentation options for tomorrow's technocrats and civil servants.

As described earlier, Nobel laureate Richard Thaler and former White House regulatory "czar" Cass Sunstein, who coauthored *Nudge* and coined the "choice architecture" locution, have thoughtfully outlined policy parameters for recommender-rich worlds. Their arguments for nudges that support "libertarian paternalism," for example—the idea that it is both possible and legitimate for private and public institutions to affect behavior while also respecting freedom of choice—clearly seek to strike a healthy balance between individual empowerment and institutional prerogative.

"Transparency and public scrutiny are important safeguards, especially when public officials are responsible for nudges and choice architecture," Sunstein maintains. "Nothing should be hidden or covert."[8] If agency matters, then informed consent becomes a technocratic obligation and ethical imperative.

Identifying policies and practices that reliably harmonize supra-self systems and user trust remain challenging. What rules and regulations around recommender transparency should be compulsory? Require trustworthy recommenders to computationally prove they do more to empower users more than to hack them? Ironically but unsurprisingly, the more data and analytics recommender systems generate, the more likely regulatory recommenders can emerge as influential public policy resources. *Pace* projections of human hackability, dystopia need not be destiny.

So if recommendation engine successes—via future regulatory and market innovation—emphasize empowerment over exploitation, what new insights and outcomes might result for users? How might recommendation innovation disrupt or redefine the Tencents, Facebooks, LinkedIns, and Amazons? Even simple technological extrapolations suggest a wealth of high-impact opportunities for driving agency, discovery, and choice. The Netflix/ByteDance "Recommendation First/ Everything Is a Recommendation" sensibility has swiftly become a dominant digital design paradigm.

Consider, as a thought experiment, a recommender based on motivational speaker Jim Rohn's provocative *bon mot* that "we are the average of the five people we spend the most time with." Whether or not you agree, take that "who we are" heuristic literally and seriously. Now digitally scrape the social media networks, shopping services, searches, and feeds they access—Instagram, Facebook, Netflix, Twitter, Twitch, Hulu, Google, LinkedIn, Amazon, Kindle, Spotify, Yelp—of your top time-intensive five for your personalized Rohn recommender. If this reminds you of hybrid recommenders blending content and collaborative filtering algorithms, you're spot on.

Do the math: the Rohn would surely deliver reasonable—if somewhat familiar—recommendations for videos to see, music to hear, places to eat, photos to share, people to meet, games to play, and things to buy. These are, after all, people you spend a lot of time with. But maybe you like and enjoy two of them more than the others. So you weight their contribution to the Rohn more heavily. That tilts the odds to get you above average quality recommendations and discoveries.

Then epiphany hits: You don't want to be the average of the five people you spend the most time with, *you want to be the average of the five best people you know*. That's your thoughtful and intentional bet on self-discovery and self-improvement. So you data scrape and "recommenderize" once more. The suggested videos, music, books, photos, and places to go truly are different from your average. You see choices you barely recognize and never thought about. But these are the best people you know and some of those suggestions look awesome.

But wait! Why limit yourself to people you know? Why not draw inspiration, motivation, and insight from accomplished people you don't know? There's plenty of celebrity information and data, for example, to scrape, process, and package. Perhaps *you want to be the average of your five favorite celebrities*. A Netflix or Hulu could easily identify the five actors, directors, or screenwriters you most enjoy and recommend *their* favorite movies and television episodes for you. A Spotify lets users compile playlists of favorite artists' playlists.

Or, more proactively, try a "meal prep" recommender; your next kitchen creations could be culled and collated from the five chefs and restaurateurs you admire. The recommendations are tailored to your skills, time, equipment, and desire. You not only see videos explaining how the dishes can be prepared, you can download augmented reality software that visually guides and talks you—via tablets and airpods—through your own preparation. The recommender (of course) connects you to your Amazon Prime or other grocery account to recommend the appropriate ingredients be ordered and/or delivered. Recommender systems evolution and technical innovation continually blur the lines between recommendation, encouragement, coaching, advice, instruction, and training. These kinds of computational mash-up scould similarly monitor and motivate your workout routines and how you parent your children. (*Do you want to be the average of the five best single moms or dads you know?*)

As machine learning and content generation capabilities exponentially increase, rethink the reach of potential recommender resources. Look to the global corpus of wisdom for inspiration and insight. *Maybe you want to be the average of the five greatest "self-help" gurus.* Consider recommendations that mash-up history's wisest—and pithiest—advisors: Epictetus, Confucius, Montaigne, Dale Carnegie, Oprah Winfrey, and Stephen Covey. Perhaps toss in Benjamin Franklin and Sigmund Freud. You want the best of their combined advice. There's no shortage of (rich) material by and about them. Scrape and "recommenderize" it.

Recurrent neural nets, topic analyzers, and natural language programming software could seamlessly stitch the combined advice into motivational narratives. Maybe Montaigne in the style of Stephen Covey enjoys the greatest influence; perhaps Epictetus in the manner of Dale Carnegie—with a dash of Viktor Frankl—truly inspires. Based on your reactions, the recommender would rank and relate from whom you should be getting advice—yet another recommended pathway to greater self-awareness and a better self.

But perhaps you want more specific recommendations. Maybe the advice you need goes beyond addressing specific skills like cooking or circumstances requiring motivation. Why not recommendation engines emphasizing particular personal attributes? *What five aspects and elements of your behavior and personality need nudging?* That is, how might recommendation and advice change when you seek to tweak parts of your self-image? For example, what books might Amazon recommend for the "curious" you versus the "typical" you? What playlist might Spotify compile for the "be more productive" you? What Power-Point and presentational imagery might be recommended for the "boldly creative" or "be more influential" you?

The point and purpose is not slavish adherence to these suggested enhancements; it's empowering people to literally see, hear, and feel what possible versions of themselves could do. In this respect, the real

The point and purpose is not slavish adherence to these suggested enhancements; it's empowering people to literally see, hear, and feel what possible versions of themselves could do.

AI behind the recommender future isn't artificial intelligence but augmented introspection. Recommendation really does become a magic mirror reflecting imaginable—imaginative?—future selves.

With tongue-in-cheek apologies to Rohn, while the "number 5" leitmotif for these recommendation gedankenexperiments is arbitrary; the essential takeaway is not. The power to personalize a portfolio of recommenders is transformative. Recommenders can be cultivated, aggregated, and personalized from individuals and communities alike. How do you want them to empower agency, insight, and choice? The more you use them, the better they learn. Recommenders algorithmically navigate the critical connections between "what to do next" and "who you want to become."

That's true whether the structure is a "one recommendation engine to rule them all" meta-ensemble or the How will you transcend today's "average"?

An important duality emerges: in these scenarios, recommenders aren't just media and mechanisms for improving consumption—the videos we watch, the music we hear, the books we read, the trips we take—they're platforms and springboards for boosting personal productivity—the presentations we make, the memos and code we write, the projects we manage, the people we motivate. That is, recommendation engines are as much about cultivating and enhancing human capital as consuming goods and services. Consequently, the key to understanding the recommender future goes beyond understanding the future of the self; it requires understanding that the rise of recommendation engines creates a future of multiple selves. Indeed, recommenders become a way to replicate and extend those aspects of ourselves that matter most.

Agency, Not Agents

Radically rethinking the capabilities of the digital self is the disruptive opportunity for dramatically empowering the human self. While agents like Alexa and Siri perform tasks to deliver desired outcomes, digital selves will actually define those tasks and outcomes. Consequently, the profound technical challenge lies less with building better agents than with enabling people to build more productive and more valuable versions of themselves.

Data-driven multiple selves, not just software swarms of agents and bots, accelerate enterprise productivity growth. And as technologies advance, they'll increasingly help people identify, manage, and measurably improve their best selves. In this future, individuals will digitally define and deploy more creative, innovative, insightful, or collaborative versions of themselves to bring new value and efficiencies to business outcomes.

They'll receive data-driven and algorithmically informed hints, nudges, and recommendations designed to align better selves with better results. Instead of consumption-driven recommenders, "multiple selfers" obtain actionable insights and advice on what to say, when to speak up, with whom to work, and how best to behave both in the moment and beyond. Tomorrow's most effective managers will employ the most effective selves. They are leaders committed to "selves" improvement.

A multiple self is best defined as a digital version of the self with one or more personal dimensions deliberately designed to significantly outperform one's ordinary, typical, or average self. The idea is to digitally amplify or enhance specific personal attributes that generate disproportionate economic impact and organizational influence. Those attributes can be affective qualities like boldness or friendliness, or technical skills such as facilitation and formulating hypotheses.

Multiple-selves management will balance the economic benefits and trade-offs between effective and affective selves. Multiple selves become innovative platforms for metacognition—thinking about how we think. Or, more precisely, they make recommendations for thinking about thinking.

Think, for example, about managing a Pareto selves portfolio—that is, what are those 20 percent chunks of talent, temperament, and behaviors that account for 80 percent of our impact, influence, and value? Not only would this "selves knowledge" be empowering, it would invite new ways for managers and leaders to motivate and measure workers. My bet is this will become a high-performance norm sooner than we expect.

As global workforces confront more agile and adaptive competitors, more traditional competencies and typical or ordinary personal performance growth no longer suffice. Research indicates that digitally deconstructing the self—grasping which aspects and attributes to amplify and what weaknesses to mitigate—unlocks a wealth of high-impact productivity opportunities.

For example:

• An executive recognizes his written communications lack clarity, energy, and forcefulness. In a multiple-selves world, the executive shares his missives and messages with software like IBM's Watson tone analyzer. The software proposes revisions, bringing force and focus to the prose.

• A global project manager seeks to encourage greater cooperation, collaboration, and esprit within her team. Her customized "selvesware" (self-analysis software) performs social-network analyses, prioritizes project milestones, and reviews post-meeting communications to propose a daily facilitative checklist.

- A technically competent but uninspired user-interface designer wants to be and be seen as more boldly creative. Specially designed visual recommenders offer prototype imagery and wire-frames based on those dimensions of creative and/or bold UX design.

In each use case, no right answer or normative solution exists; but individuals get clear, compelling, and customized choices they wouldn't otherwise have. As with Amazon, Google Maps, and Netflix, people receive actionable, data-driven recommendations informed by algorithms explicitly designed to create a desired self.

Whether Amazon/Netflix-style recommendations, Atul Gawande *Checklist Manifesto*–like checklists, or brave new genres of behavioral nudges best facilitate selves improvement is an affective/cognitive question driving human-capital research and development. Tomorrow's selves-motivated workers digitally choose to invest in who they need to become, not just what they're supposed to do next.

The social-science research and concepts for animating multiple selves is impressively robust. The multiple-selves future draws from a rich—and growing—reservoir of psychology, behavioral economics, and cognitive research into how people actually make productive choices.

The extensive literature essentially observes that the human mind is less a coherent whole than a clash of competing cognitive perspectives and affective desires. The self—or human agency—is both product and byproduct of those inherent and perennial conflicts.

"To understand the most important ideas in psychology," observes New York University research psychologist, Jonathan Haidt, "you need to understand how the mind is divided into parts that sometimes conflict. We assume there is one person in each body, but in some ways we are each more like a committee whose members have been thrown together working at cross-purposes."[9]

Let technology turn the apparent "bug" of a divided self into the productive feature of digital selves. Treat the vast research literature on the self as a resource for recasting that division into mindful and purposeful outcomes. As mentioned, Daniel Kahneman's Nobel Prize–winning research defining cognitive biases, heuristics, and prospect theory proffers clear frameworks for designing digital selves, while behavioral economics—with its empirically demonstrated insights into anchoring, framing, and hyperbolic discounting—paves the way for selvesware that promotes enhanced self-awareness. Similarly, the research of Nobel laureate economists Tom Schelling (egonomics) and Herbert Simon (bounded rationality and satisficing) serve as conceptual inspiration for technical instantiation.

MIT AI pioneer Marvin Minsky's *Society of Mind*, for example, offers a veritable roadmap for researchers and entrepreneurs seeking insights into what modules of the mind are best positioned for digital augmentation and enhancement. The multiple-selves thesis enjoys remarkably multidisciplinary support. All of these works strongly suggest that cultivating and managing multiple selves empower personal productivity. While that doesn't make software agents less valuable or important, it persuasively suggests that human agency's productive potential is underappreciated.

Ongoing global trends make this new human capital investment option appealing as well. Widespread adoption of "quantified-self" tools and technologies—think wearable devices and sensors like Fitbit and the Apple Watch—promise ever-richer datasets for multiple-selves design. Technologies that track steps and heart rates already draw actionable inferences about individual energy levels and mood. Mobile device apps can easily play significant workplace roles in assessing mental acuity and attention just as they do for physical fitness.

The workday is near when selvesware instrumentation and personal KPI dashboards can physiologically sense when users are not in the mood to take advice, respond to recommendations or "Slack chat" the boss. The results? More granular self-data and analytics will prove essential ingredients for boosting personal productivity and performance. Innovative, curious, facilitative, communicative, and other value-added/value-adding selves will get the right cues, nudges, and recommendations at the right moments prompted by increasingly sophisticated selvesware.

The business case for this recommender future remains simple and straightforward: well-managed multiple selves will reliably out-perform and out-produce average selves assisted by agents and bots.

The global trend to workplace analytics that both complement and reinforce the quantified-selves capability highlights this disruptive design. High-performance companies and cultures like Google and China's Heier also portend the personal productivity future. Laszlo Bock's excellent book *Work Rules!* captures how thoughtfully his former company invested in data collection and metrics for assessing team, as well as individual, value. To test managerial assumptions, the company conducts almost as many experiments inside the organization as out. And Google's dedication to relentless improvement, Bock writes, makes it "open to crazy ideas."[10]

Born-digital companies and cultures like Google, Facebook, Amazon, Booking.com and Spotify—with their digital sophistication and algorithmic chops—are supremely well positioned to acquire further competitive advantage by cultivating workforces of high-performance multiple-selves. As enterprise networks grow, so do the options. A facilitative multiple self, for instance, may search Google, Bing, or LinkedIn to gather data on potential collaborators. A creative multiple self might say, "Alexa, find me an image with this aesthetic sensibility," or "Siri, find a creative self in the company that best complements me."

In effect, productive humans won't just manage portfolios of software agents, but teams of multiple selves. In turn, managers oversee not just teams of individuals, but networked ensembles of multiple selves and agents as well.

In theory, a multiple selves as a service—MSAAS—provider could charge people $199 per year per self, or 1 percent of gross income—whichever is less—to sign on. People and their employers can quickly decide if "better selves" are worth the money compared to say, an Amazon Prime, Spotify, or Netflix subscription.

More seriously, the real business challenge has less to do with business models around recommender systems or machine learning technology than data governance. How should companies facilitate data access and sharing of personnel data and workplace analytics? Does your "boldly creative self" at work belong to your employer or to you? How portable or proprietary should one's data-driven productive selves be?

UCLA psychologist Hal Hershfield has conducted a series of powerful and persuasive experiments crystallizing how important the potential of future selves can be to the current moment. Though rooted in sophisticated visualization technology, the underlying hypothesis is shockingly simple and compelling: seeing one's future self measurably influences current decision.

In one study, Hershfield showed some college students images of their own faces that had been digitally altered to appear fifty years older. He showed others their contemporary images. While looking at their photos, participants were asked to indicate how much of their salary they wanted to allocate to their 401(k) retirement accounts. Those who'd seen a glimpse of their digitally aged selves said that they would save about 30 percent more, on average, than the students shown pictures of their current selves.

In a related experiment, participants encouraged to think about their selves twenty years hence committed to exercising more today. A field study of Mexican consumers found people prompted to think about their future selves were more likely to sign up for an automated savings account than participants not similarly nudged; the control group had a 1 percent sign-up rate, while the "future selves" group sign-ups approached 3 percent. "So the lesson from that is," Hirshfield said, "anything that we can do that will increase how concrete and salient our future self is—that's the type of thing that can help us make better decisions."[11]

As recommender systems and selvesware technologies improve and converge, individuals will have the opportunity and option to forge similar connections to images and representations of (possible) future selves. Envisioning how a physically fit future self looks is easy. But how best to conjure and visualize one's creative or influential future self? Seeing one's current options—experiencing in-the-moment nudges and advice—is not enough. The challenge these breakthroughs pose is how best to help people connect with their possible future selves. Can recommenders and selvesware combine to help users better bridge "what to do next" and "who they want to become"? Trend may not be destiny, but it seems clear that innovative opportunities for agency and empowerment more than rival the real risks of dependency and exploitation.

"Know thyself" is one of the oldest and wisest of classic aphorisms and advice. Its intrinsic wisdom has deservedly stood the test of time. The rise of ever-smarter, ever-more-disruptive technologies, however, strongly recommends a twenty-first-century digital update: Know thy selves.

ACKNOWLEDGMENTS

Before writing these acknowledgments, I took a moment to reread my acknowledgments from books past. They appropriately reflected both my genuine gratitude and overwhelming relief. Acknowledgments here require an additional attribute: humility.

I found the challenge of capturing, distilling, communicating, and truly understanding the "essential knowledge" of recommendation engine mathematics and technology during a time of disruptive transformation humbling. This book would have failed without the insightful generosity—and generous insight—of people who took time and care not just to help me but to help inform my readers. You.

The best of this book builds on their thoughtful contributions, comments, and review. Any mistakes, misinterpretations, or errors are mine and mine alone. That said, this book was exciting to write. Its essential themes are globally important, timely, and destined to enormously influence people's perceived—and real—freedoms.

Emily Taber, my editor, has been my primary and essential partner for this book. Her comments, critiques, responsiveness, and writer-whisperer style combined to make the process less painful and more productive. I am grateful beyond words (irony intended). This would not have been possible without her. I am also grateful to and for Gita Manaktala, a dear friend who is the Editorial Director at the MIT Press. She's a remarkable person and editor; I'm sure she effectively protected myself, Emily, and this book from my lesser qualities.

MIT Press's Michael Sims ably copyedited this book and procured Tyler Mayo, a careful researcher, to check facts and attributions. Deborah Cantor-Adams has skillfully completed the editing of this manuscript and guided it through production.

My MIT colleagues continue to be enormously supportive. Special thanks to the MIT Initiative on the Digital Economy's Erik Brynjolfsson and Andrew McAfee—whose research and thought-leadership is world-class—along with David Verrill and Christine Ko. They are friends as well as colleagues. IDE's Paula Klein, whose interviews and editing sharpened my research communications, was particularly helpful. Marshall van Alstyne and Geoff Parker, who run MIT's pioneering Platform Summit, have also been terrific.

As my work on recommender systems and "selvesware" evolved, I had increasing opportunities to work more closely with *MIT Sloan Management Review*. My ongoing collaborations with David Kiron and Allison Ryder have proven remarkably productive. I'm grateful to them.

Harvard Business Review's Melinda Merino has also been a source of incisive editorial support.

Between my executive education classes and MIT's Industrial Liaison Program, I enjoyed multiple venues for publicly testing and refining my future of agency, advice, and recommendation hypotheses. My students combine intellectual curiosity with an intense desire to make things work in the real world. My ILP colleagues—particularly Karl Koster, Todd Glickman, Tony Knopp, Marie van der Sande, Randall Wright, Sheri Brodeur, Jewan Bae, Corey Cheng, Daphne de Baritault, Marie DiCicco, Hong Fan, Ken Goldman, Peter Lohse, Rachel Obera-Soltz, Steven Palmer, Erik Vogan, Graham Rong, Klaus Schleicher, Ron Spangler, and Irina Sigalovsky—all introduced me to people and organizations who proved to be superior sounding boards and resources for testing my book's themes in real-world contexts.

I am also grateful to Tom Malone, Ronni Kohavi, Greg Linden, Patrick Hebron, Brad Klingenberg, and assorted friends and colleagues from Amazon, Facebook, Alibaba, Microsoft, Netflix, booking.com, and Google for taking the time to both answer questions and think out loud about the recommendation future.

With not a hint of embarrassment or self-consciousness, I'd like to conclude by acknowledging the authors of several Essential Knowledge books—notably

Free Will, *Machine Learning*, *Data Science*, *Metadata*, and *Self-Tracking*—for their inspirations and insights. Their hard work informed my own. I hope this contribution to the series meets the standard I aspired to.

For a variety of reasons, I am particularly grateful to my wife, Beth Ann. She is a wonderful recommendation engine that I ignore at my peril. Her non-editorial contributions make this book and my life possible.

Association rules
Conditional "if-then" statements asserting the likelihood or probability of relationships between data items within large data sets in a variety of related databases.

Autoencoder
A type of artificial neural network used to produce efficient representations of data in an unsupervised and nonlinear manner.

Bayes's theorem
A famous theorem used by statisticians to describe the probability of an event based on prior knowledge of conditions that might be related to an occurrence.

Choice architecture
The practice of influencing choice by organizing the context in which people make decisions. They describe how decisions are affected by the layout, sequencing and/or range of available choices.

Classification
The process of predicting the class of given data points. Classes are sometimes called as targets/ labels or categories.

Cluster
A collection of data points aggregated together because of certain similarities; the data seem to be "gathered" around a particular value.

Cold start
A term used to describe the circumstance that the recommendation engine cannot draw significant inferences for either users or items because it doesn't have enough information yet

Collaborative filtering
A method used in the context of recommendation engines to make predictions about the interests of a user by collecting preferences from a larger group

of users. Collaborative filtering is based on the assumption that people who agreed in the past will agree in the future and that they will like similar kinds of items as they liked in the past.

Content-based recommenders
Algorithms that measure similarity by looking for common features of the items to be recommended.

Curse of dimensionality
Phenomena that arise when analyzing and organizing data in high-dimensional spaces because the greater the number of dimensions, the sparser the amount of relevant data becomes.

Demographic recommender
Provides recommendations based on a demographic profile of the user.

Dimensionality reduction
The process of reducing the number of random variables under consideration by obtaining a set of principal/key variables.

Diversification
Identifying and recommending items dissimilar from each other but contextually relevant to the user's interests.

Embedding
A mathematical mapping from discrete objects, such as words, to vectors of real numbers. More explicitly, it is a relatively low-dimensional space into which high-dimensional vectors are translated. Embeddings make machine learning on large inputs like sparse vectors representing words easier.

Ensemble methods
In statistics and machine learning, ensemble methods use multiple learning algorithms to obtain better predictive performance that could be obtained from any of the constituent learning algorithms alone.

Evaluation metrics
Used to determine the quality of the recommendation engine's recommendation.

Explicit interactions
Human/machine interactions where users intentionally rate, rank, review, or otherwise offer deliberate feedback to a proffered option.

"Explore vs. exploit" trade-off
The computation to determine whether exploring—gathering more information that might lead to better future decisions or exploitation—making the best decision given current information—represents the better option.

Feature (feature selection)
A variable that is used as an input to a model.

Feature vector
A set of values (a vector) able to describe an object (an image or product, for example) that measurably differentiate it from other objects (images or products).

Filter bubble
Recommendation and personalization systems risk producing around us a set of information limited and close to the user's interests, closing it in a sort of information bubble that excludes new ideas or other important information.

Hybrid recommendation engine
Any recommendation engine that combines multiple recommendation techniques together to produce its output, for example a collaborative filter recommender and a content-based recommender.

Implicit interactions
Human/machine interactions that capture indirect "behaviorial byproducts" of users and use. For example, time spent viewing a video or clicking/swiping an onscreen offer. Implicit interaction systems seek to understand user intentions by capturing user actions in context.

Inference
The process of making predictions by applying a trained model to new, unlabeled instances.

Item profiles
These consist of features of items. Different kinds of items have different features on which content-based similarity can be based. Features of documents, for example, are typically important or unusual words.

Item-based collaborative filtering
Measures the similarity between the items that target users rates/ interacts with and other items.

Knowledge-based recommender
Provides recommendations of items based on specific inferences about user needs and preferences.

Latent features
Variables that are not directly observed but are rather inferred through a mathematical model from other variables that are directly measured or observed. Latent features are computed from observed features using matrix factorization.

Learning-to-rank
The application of machine learning to the construction of ranking models for recommender systems.

Market basket analysis
An analytic technique for identifying the strength of association between pairs of products typically, frequently, or occasionally purchased together. These analytics identify patterns of co-occurrence.

Matrix factorization (also called decomposition)
A way of reducing a larger matrix into its constituent parts. The approach simplifies more complex matrix operations that can be performed on the decomposed matrix rather than on the original matrix itself. It is used to discover latent—or hidden—features between two entities.

Matrix
A rectangular array of numbers, symbols, or expressions arranged in rows and columns.

Memory-based recommenders
Algorithms that compute the similarities in memory without the need to produce a model first.

Model-based recommenders
Algorithms that use the ratings and interaction data to learn a predictive model, which is used to recommend new or unrated items.

Multiarmed bandits
A term used to describe a problem in which a fixed limited set of resources must be allocated between competing choices in a way that maximizes their expected gain for the player.

Nearest-neighbor
A data classification approach that attempts to determine what group a data point should belong in by looking at the data points around it.

Recommendation engines
Algorithms that attempt to predict a user's response to an item by discovering similar items and the response of the user to those.

Regression
A statistical measurement that attempts to determine the strength of the relationship—the correlation—between a dependent variable and a series of other changing—independent—variables.

Reinforcement learning
Learning that occurs through interaction with an environment. Reinforcement learning is essentially *trial and error* learning tied to a reward function. A "reinforcement learning" agent learns from the consequences of its actions, rather than from being explicitly taught or trained. It selects its next actions on the basis of its past experiences (exploitation) and also by new choices (exploration).

Root mean square error
a measure of how well the model performed; it measures the difference between predicted values and the actual values.

Serendipity
Unexpectedly relevant and appealing recommendation.

Similarity measure or similarity function
Quantifies the similarity between two objects. That is, they calculate how "close" or "alike" objects are in mathematical sets or dimensions. Similarity is often computed as the inverse of distance metrics; that is, how far apart items are in mathematical space.

Singular value decomposition
A mathematical technique for deconstructing a complicated matrix into smaller matrixes in ways that make computation faster, simpler and more reliable. These computations can detect and identify unexpectedly relevant groupings.

Social graph
A diagram of the links and interconnects between and among people, groups and organizations in a social network. The term also describe an individual's social network.

Sparsity
The relative absence of relevant data for identifying "neighbors" or computing statistically significant similarities. Sparsity limits recommender quality and applicability of collaborative filtering techniques.

TF-IDF (term frequency-inverse document frequency)
A numerical statistic intended to reflect how important a word is to particular documents in a collection or body of work.

Topic modeling
A category of unsupervised machine learning algorithms that uses clustering to find hidden structures in textual data and interpret them as topics.

NOTES

Introduction
1. Michael Schrage, "Recommendation Nation," *MIT Technology Review*, April 22, 2008. https://www.technologyreview.com/s/409956/recommendation-nation.

Chapter 1
1. "Recommender System," last modified 14 July 2019. https://en.wikipedia.org/wiki/Recommender_system.

2. Francesco Ricci, Lior Rokach, Bracha Shapira, and Paul B. Kanto, eds., *Recommender Systems Handbook* (Springer, 2011).

3. Anders Lindstrom "Intelligent and Relevant," master's thesis, KTH Royal Institute of Technology, Stockholm, 2008.

4. https://www.quora.com/What-are-some-of-the-interesting-innovative-use-cases-in-recommendation-systems.

5. https://www.tubefilter.com/2019/05/07/number-hours-video-uploaded-to-youtube-per-minute.

6. Yan Yan, Wentao Guo, Meng Zhao, Jinghe Hu, and Weipeng P. Yan, "Optimizing Gross Merchandise Volume via DNN-MAB Dynamic Ranking Paradigm," Cornell University arXiv:1708.03993.

7. https://www.exastax.com/recommendation-systems/driving-revenue-with-personalized-product-recommendations.

8. Alibaba.com.

9. J. Ben Schafer, Joseph Konstan, and John Riedl, "Recommender Systems in E-Commerce," *EC '99: Proceedings of the 1st ACM Conference on Electronic Commerce, Denver, CO, November 3-5, 1999*, 158–166 (New York: ACM).

10. J. Harmon, "Dynamics of Human Trust in Recommender Systems," *Proceedings of the 8th ACM Conference on Recommender Systems*, 2014. https://www.researchgate.net/publication/274899847_Dynamics_of_human_trust_in_recommender_systems.

11. Personal 2015 interviews with Sales Predict cofounder and CEO Yaron Zakia-Or.

12. Kimiz Dalkir, "Measuring the Impact of Social Media," in *Social Knowledge: Using Social Media to Know What You Know*, ed. John P. Girard and JoAnn L. Girard (Hershey, PA: Information Science Reference, 2011), 35.

13. Jack Herlocker, Joseph A. Konstan, Loren G. Terveen, and John T. Riedl, "Evaluating Collaborative Filtering Recommender Systems," *ACM Transactions on Information Systems* 22, no. 1 (January 2004).

14. https://www.forbes.com/sites/bernardmarr/2018/05/21/how-much-data-do-we-create-every-day-the-mind-blowing-stats-everyone-should-read/#61b4820160ba.

15. https://www.forbes.com/sites/bernardmarr/2018/05/21/how-much-data-do-we-create-every-day-the-mind-blowing-stats-everyone-should-read/#61b4820160ba.

16. https://venturebeat.com/2017/09/12/kai-fu-lee-talks-ai-driven-unemployment-who-says-we-need-jobs.

17. http://radar.oreilly.com/2006/05/my-commencement-speech-at-sims.html.

18. http://glinden.blogspot.com/2006/05/tim-oreilly-and-defining-web-20.html.

19. Quoted in https://fs.blog/2017/07/filter-bubbles.

20. https://www.linkedin.com/pulse/taste-trust-daniel-tunkelang.

21. https://hbr.org/2013/01/jeff-bezos-on-leading-for-the.

Chapter 2

1. Cicero, *De Divinatione* 1.2, *Ancient Greek Divination*, trans. Sarah Iles Johnston (Chichester: Wiley-Blackwell, 2008), 3.

2. Jamie Fisher, "Peter Struck's Odyssey," *Pennsylvania Gazette*, April 24, 2017. http://thepenngazette.com/peter-strucks-odyssey.

3. Cicero, *De Divinatione* 5.11, cited in Clifford Pickover, *Dreaming the Future: The Fantastic Story of Prediction* (Amherst, NY: Prometheus Books, 2001), 24.

4. Pickover, *Dreaming the Future*.

5. Cicero, *De Divinatione* 1.2.

6. Peter Struck, "A World Full of Signs: Understanding Divination in Ancient Stoicism," in *Seeing with Different Eyes: Essays in Astrology and Divination*, ed. Patrick Curry and Angela Voss (Newcastle: Cambridge Scholars Press, 2008), 3.

7. Peter Struck, *Divination and Human Nature: A Cognitive History of Intuition in Classical Antiquity* (Princeton: Princeton University Press, 2016), 11.

8. Struck.

9. Jim Tester, *A History of Western Astrology* (Suffolk: The Boydell Press, 1987), 11.

10. Ian Bacon, "The Other 'A' Word—Astrology in the Classical World." https://astronomy.swin.edu.au/sao/guest/bacon.

11. Henry Morley, *Jerome Cardan: The Life of Girolamo Cardano, of Milan, Physician* (London: Chapman and Hall, 1854), 119–120.

12. Walter William Rouse Ball, *A Short Account of the History of Mathematics* (London: Macmillan and Co., 1908), 222.

13. Philip J. Davis, *Mathematics & Common Sense: A Case of Creative Tension* (Natick, MA: A. K. Peters, 2006), 197.

14. Anthony Grafton, *Cardano's Cosmos: The Worlds and Works of a Renaissance Astrologer* (Cambridge, MA: Harvard University Press, 1999), 10.

15. Pierre-Simon Laplace, *Philosophical Essay on Probabilities*, trans. Andrew Dale (New York: Springer Verlag, 1995), 107.

16. Richard Mussard, "Wisdom Literature and the Quest for Wisdom: Babylon and Beyond." https://www.usi.edu/media/2431744/ram-mussard.pdf, 2005, 2.

17. Harold Bloom, *Where Shall Wisdom Be Found?* (New York: Riverhead Books, 2004), 1, cited in Mussard, "Wisdom Literature and the Quest for Wisdom."

18. Aelius Theon, *Prog.* 1.

19. Marshall McLuhan, *Understanding Me: Lectures and Interviews*, ed. Stephanie McLuhan and David Staines (Cambridge, MA: MIT Press, 2003), 61.

20. Elizabeth Eisenstein, "Some Conjectures about the Impact of Printing on Western Society and Thought: A Preliminary Report," *Journal of Modern History* 40 (1968): 8.

21. Eisenstein, 41.

22. Eisenstein, 42.

23. William Irvine, *A Guide to the Good Life: The Ancient Art of Stoic Joy* (New York: Oxford University Press, 2009), 226.

24. Michael Bhakskar, "In the Age of the Algorithm, the Human Gatekeeper Is Back," *The Guardian*, September 30, 2016. https://www.theguardian.com/technology/2016/sep/30/age-of-algorithm-human-gatekeeper.

Chapter 3

1. David Goldberg, David Nichols, Brian M. Oki, and Douglas Terry, "Using Collaborative Filtering to Weave an Information Tapestry," *Communications of the ACM.* 35, no. 12 (1992): 61–70.

2. John Riedl and Joseph Konstan, *Word of Mouse: The Marketing Power of Collaborative Filtering* (New York: Warner Books, 2002).

3. Paul Resnick and Hal R. Varian, "Recommender Systems," *Communications of the ACM* 40, no. 3 (1997): 56–58.

4. Dietmaar Jannach and Gerhard Friedrich, "Tutorial: Recommender System." http://u.cs.biu.ac.il/~sarit/advai2015/recommendation-short-2017.pdf.

5. "Video Interview with Brad Miller," Coursera.org.https://www.coursera.org/lecture/collaborative-filtering/interview-with-brad-miller-AZKVP.

6. Greg Linden, "Early Amazon: Similarities," March 22, 2006. http://glinden.blogspot.com/2006/03/early-amazon-similarities.html.

7. Greg Linden, "Early Amazon: Shopping Cart Recommendations," April 25, 2006. http://glinden.blogspot.com/2006/04/early-amazon-shopping-cart.html.

8. https://www.inc.com/jessica-stillman/7-jeff-bezos-quotes-that-will-make-you-rethink-success.html.

9. Linden "Early Amazon: Shopping Cart Recommendations

10. Olav Sorenson, "Netflix," 2004. http://www.olavsorenson.net/wp-content/uploads/2013/06/Netflix.pdf.

11. Jeffrey M. O'Brien, "The Netflix Effect," *Wired*, December 1, 2012. https://www.wired.com/2002/12/netflix-6.

12. Scott Page, "The Netflix Prize." https://mellon.org/initiatives/our-compelling-interests/excerpts/netflix-prize.

13. Eliot Van Buskirk, "How the Netflix Prize Was Won," *Wired*, September 22, 2009. https://www.wired.com/2009/09/how-the-netflix-prize-was-won.

14. Joseph Konstan, "How Important Was the Netflix Prize for the Recommender Systems Area?." https://www.quora.com/How-important-was-the-Netflix-Prize-for-the-Recommender-Systems-area.

15. Tom Vanderbilt, "The Science behind the Netflix Algorithms That Decide What You'll Watch Next," *Wired*. August 7, 2013. https://www.wired.com/2013/08/qq-netflix-algorithm.

16. Vanderbilt.

17. David Carr, "Giving Viewers What They Want," *New York Times*. February 25, 2013. https://www.nytimes.com/2013/02/25/business/media/for-house-of-cards-using-big-data-to-guarantee-its-popularity.html.

18. John Ciancutti, "Does Netflix Add Content Based on Your Searches?" https://www.quora.com/Netflix-product/Does-Netflix-add-content-based-on-your-searches/answer/John-Ciancutti.

19. James Davidson, Benjamin Liebald, Junning Liu, Palash Nandy, and Taylor Van Nest, "The YouTube Recommendation System," *RecSys* 2010. https://www.inf.unibz.it/~ricci/ISR/papers/p293-davidson.pdf, 294.

20. Davidson, Liebald, Liu, Nandy, and Van Nest, 295.

21. Shubhi Tandon, "YouTube's Cristos Goodrow on Why Views Are a 'Bad' Metric," *Digital Markets Asia*, September 23, 2015. http://www.digitalmarket.asia/youtubes-cristos-goodrow-on-why-views-are-a-bad-metric.

22. David Gelles, "Inside Match.com," *Financial Times,* July 29, 2011. https://www.ft.com/content/f31cae04-b8ca-11e0-8206-00144feabdc0.

23. John Cacioppo, Stephanie Cacioppo, Gian C. Gonzaga, Elizabeth L. Ogburn, and Tyler VanderWeele, "Marital Satisfaction and Break-Ups Differ across On-line and Off-line Meeting Venues," *PNAS* 110, no. 25 (2013): 10135–10140.

24. Steven Levy, "How Google Is Remaking Itself as a "Machine Learning First" Company," *Wired*, June 26, 2016. https://www.wired.com/2016/06/how-google-is-remaking-itself-as-a-machine-learning-first-company.

25. Casey Newton, "How YouTube Perfected the Feed," *The Verge*, August 30, 2017. https://www.theverge.com/2017/8/30/16222850/youtube-google-brain-algorithm-video-recommendation-personalized-feed.

26. Newton.

Chapter 4

1. Christian Rudder, *Dataclysm: Love, Sex, Race, and Identity—What Our On-line Lives Tell Us about Our Offline Selves* (New York: Random House, 2014), 19–20.

2. Amit Sharma, "What Are Some of the Interesting/Innovative Use Cases in Recommendation Systems?" https://www.quora.com/What-are-some-of-the-interesting-innovative-use-cases-in-recommendation-systems.

3. Joseph Konstan and John Riedl, "Deconstructing Recommender Systems," *IEEE Spectrum*, August 30, 2017. https://spectrum.ieee.org/computing/software/deconstructing-recommender-systems.

4. Rudder, *Dataclysm.*

5. Petros Domingos, "A Few Useful Things to Know About Machine Learning," *Communications of the ACM* 65 (10), 2012, 79.

6. Joe Davidson, "No, Machine Learning Is Not Just Glorified Statistics," *Towards Data Science*, June 27, 2018. https://towardsdatascience.com/no-machine-learning-is-not-just-glorified-statistics-26d3952234e3.

7. Stephen L. Scott, "Overview of Content Experiments: Multi-armed Bandit Experiments." https://support.google.com/analytics/answer/2844870?hl=en.

8. David Silverman, "Machines We Can Trust, Learn from and Collaborate With." https://www.imperial.ac.uk/enterprise/issues/explainable-ai.

9. E. S. Vormand and A. D. Miller, "Assessing the Value of Transparency in Recommender Systems: An End-User Perspective," *Proceedings of the 5th Joint*

Workshop on Interfaces and Human Decision Making for Recommender Systems co-located with ACM Conference on Recommender Systems 2225, 61–68. http://ceur-ws.org/Vol-2225.

10. James McInerney, "Explore, Exploit, and Explain: Personalizing Explainable Recommendations with Bandits," October 1, 2018. http://jamesmc.com/blog/2018/10/1/explore-exploit-explain.

11. Zachary Lipton, "The Mythos of Model Interpretability," *2016 IMCL Workshop on Human Interpretability in Machine Learning*, 2016. Available at https://arxiv.org/pdf/1606.03490.pdf.

Chapter 5

1. Xavier Amatriain and Justin Basilico, "Netflix Recommendations: Beyond the 5 stars," April 6, 2012. https://medium.com/netflix-techblog/netflix-recommendations-beyond-the-5-stars-part-1-55838468f429.

2. Tom Vanderbilt, "The Science behind the Netflix Algorithms That Decide What You'll Watch Next," *Wired*. August 7, 2013. https://www.wired.com/2013/08/qq-netflix-algorithm.

3. Amatriain and Basilico, "Netflix Recommendations."

4. Deborah D'Souza, "Netflix Doesn't Want to Talk about Binge-Watching," May 18, 2019. https://www.investopedia.com/tech/netflix-obsessed-binge-watching-and-its-problem.

5. Cass Sunstein and Richard Thaler, *Nudge: Improving Decisions about Health, Wealth and Happiness* (New York: Penguin Books, 2009), 6.

6. Julian Baggini, "How Nudge Theory Is Aging Well," *Financial Times*, April 19, 2019. https://www.ft.com/content/4271cfac-5a26-11e9-840c-530737425559.

7. Della Bradshaw, "How a Little Nudge Can Lead to Better Decisions," *Financial Times*, November 25, 2015. https://www.ft.com/content/e98e2018-70ca-11e5-ad6d-f4ed76f0900a.

8. Paolo Cremonesi, Antonio Donatacci, Franca Garzotto, and Roberto Turrin , "Decision-Making in Recommender Systems: The Role of User's Goals and Bounded Resources," *RecSys* ,2012. https://pdfs.semanticscholar.org/33e1/d81fd138a5d88d6522cc56a38df230b9b0bf.pdf, 6.

9. B. J. Fogg, "A Behavior Model for Persuasive Design," *Proceedings of the Fourth International Conference on Persuasive Technology*, 2009.

10. Jordan Larson, "The Invisible, Manipulative Power of Persuasive Technology," *Pacific Standard*, May 14, 2014. https://psmag.com/environment/captology-fogg-invisible-manipulative-power-persuasive-technology-81301.

11. Nick Nelson, "The Power of a Picture," May 3, 2016. https://media.netflix.com/en/company-blog/the-power-of-a-picture.

12. Ashok Chandrashekar, Fernando Amat, Justin Basilico, and Tony Jebara, "Artwork Personalization at Netflix," December 7, 2017. https://medium.com/netflix-techblog/artwork-personalization-c589f074ad76.

13. Interview with Justin Basilico. https://qconsf.com/sf2018/presentation/artwork-personalization-netflix.

14. Mike O'Brien, "Machine Learning and Choice Paralysis: How Netflix Personalizes Title Images," December 18, 2017. https://www.clickz.com/machine-learning-choice-paralysis-netflix-personalizes-title-images/204354.

15. "How Does Netflix Know I like Black Mirror," February 11, 2018. https://www.mediaan.com/black-mirror-netflix.

16. Brent Smith and Linden, Greg. "Two Decades of Recommender Systems at Amazon.com," *IEEE Internet Computing* 21 (May–June 2017): 12–18.

Chapter 6

1. https://www.wired.co.uk/article/tastemakers-spotify-edward-newett.

2. https://qz.com/571007/the-magic-that-makes-spotifys-discover-weekly-playlists-so-damn-good.

3. https://techcrunch.com/2016/05/25/playlists-not-blogs.

4. https://www.wired.co.uk/article/tastemakers-spotify-edward-newett.

5. https://spectrum.ieee.org/view-from-the-valley/computing/software/the-little-hack-that-could-the-story-of-spotifys-discover-weekly-recommendation-engine.

6. https://996.ggvc.com/2018/01/15/episode-4-liu-zhen-on-bytedances-global-vision-and-leading-uber-china.

7. https://www.businessofapps.com/data/tik-tok-statistics.

8. Emily Feng, "Toutiao Touts AI for Individual News in Vast China Market," *Financial Times* https://www.ft.com/content/04b55fdc-28b7-11e7-9ec8-168383da43b7.

9. https://technode.com/2017/12/07/toutiao-machine-learning.

10. https://technode.com/2017/12/07/toutiao-machine-learning.

11. https://futurism.com/the-future-of-writing-chinas-ai-reporter-published-450-articles-during-rio-olympics.

12. https://technode.com/2017/12/07/toutiao-machine-learning.

13. https://www.kleinerperkins.com/perspectives/internet-trends-report-2018.

14. https://nguyenvanchuong.com/blog/2018/04/11/how-douyin-became-one-of-chinas-top-micro-video-apps-in-500-days.

15. https://foreignpolicy.com/2019/01/16/bytedance-cant-outrun-beijings-shadow.

16. https://hbr.org/2018/05/stitch-fixs-ceo-on-selling-personal-style-to-the -mass-market.

17. https://hbr.org/2018/05/stitch-fixs-ceo-on-selling-personal-style-to-the -mass-market.

18. https://digitalbusiness.substack.com/p/stitch-fix-game-collects-data-for.

19. https://www.datainnovation.org/2016/05/5-qs-for-eric-colson-chief -algorithms-officer-at-stitch-fix.

20. https://www.washingtonpost.com/business/economy/the-personal-stylists -who-are-training-the-bots-to-be-personal-stylists/2018/08/17/69bb476a-9f1d -11e8-93e3-24d1703d2a7a_story.html?utm_term=.a361c35fce59.

21. https://blog.fastforwardlabs.com/2016/05/25/human-machine-algorithms -interview-with-eric.html.

22. https://multithreaded.stitchfix.com/blog/2016/07/14/data-driven-fashion -design.

Chapter 7

1. Greg Linden, "Tech and Tech Idealism," May 8, 2019. http://glinden .blogspot.com.

2. Seth Stephens-Davidowitz, *Everybody Lies: Big Data, New Data, and What the Internet Can Tell Us About Who We Really Are* (New York: Harper Collins, 2017).

3. "Online Retail Is Booming in China," *The Economist*, October 26, 2017. https://www.economist.com/special-report/2017/10/26/online-retail-is-booming -in-china.

4. Steve Paulson, "Yuval Noah Harari Is Worried About Our Souls," *Nautilus*, December 27, 2017. http://nautil.us/issue/67/reboot/yuval-noah -harari-is-worried-about-our-souls.

5. Nicholas Thompson, "When Tech Knows You Better Than You Know Yourself," *Wired*, October 4, 2018. https://www.wired.com/story/artificial -intelligence-yuval-noah-harari-tristan-harris.

6. Yuval Noah Harari, "Yuval Noah Harari: The Myth of Freedom," *The Guardian*, September 14, 2018. https://www.theguardian.com/books/2018/sep/14/ yuval-noah-harari-the-new-threat-to-liberal-democracy.

7. Sharon Gaudin, "At Stitch Fix, Data Scientists and A.I. Become Personal Stylists," *Computerworld*, May 6, 2016. https://www.idginsiderpro .com/article/3067264/at-stitch-fix-data-scientists-and-ai-become-personal -stylists.html.

8. Cass Sunstein, *The Ethics of Influence: Government in the Age of Behavioral Science* (New York: Cambridge University Press, 2016), 201.

9. Jonathan Haidt, *The Happiness Hypothesis: Finding Modern Truth in Ancient Wisdom* (New York: Basic Books, 2006), 9.

10. Laszlo Bock, *Work Rules! Insights from Inside Google That Will Transform How You Live and Lead* (New York: Hachette, 2015), 359.

11. Melissa Dahl, "It's Time to Get Acquainted with Your Future Self," *The Cut*, January 14, 2015. https://www.thecut.com/2015/01/time-to-get-acquainted -with-future-you.html.

FURTHER READING

Aggarwal, Charu C. *Recommender Systems: The Textbook*. Springer, 2016.

Alpaydin, Ethem. *Machine Learning*. MIT Press Essential Knowledge series, 2016.

Balaguer, Balaguer. *Free Will*. MIT Press Essential Knowledge series, 2014.

Chen, Minmin. "Reinforcement Learning for Recommender Systems: A Case Study on Youtube," https://www.youtube.com/watch?v=HEqQ2_1XRTs

Falk, Kim. *Practical Recommender Systems*. Manning Publications, 2019.

Jannach, Dietmar, and Markus Zanker. *Recommender Systems: An Introduction*. Cambridge University Press 2010.

Karatzoglou, Alexandros. "Deep Learning for Recommender Systems." https://www.youtube.com/watch?v=hDjEd43R7Ik

Pickover, Clifford A. *Dreaming the Future: The Fantastic Story of Prediction*. Prometheus Books, 2001.

RecSys 2016: Tutorial on Lessons Learned from Building Real-life Recommender Systems. https://www.youtube.com/watch?v=VJOtr47V0eo

RecSys 2016: Paper Session 7—Past, Present, & Future of Recommender Systems: Industry Perspective, https://www.youtube.com/watch?v=QOiZRe0UnRw

Theobald, Oliver. *Machine Learning: Make Your Own Recommender System*. Independently published, 2018.

INDEX

The MIT Press Essential Knowledge Series

MICHAEL SCHRAGE is a Research Fellow at the MIT Sloan School of Management's Initiative on the Digital Economy. A sought-after expert on innovation, metrics, and network effects, he is the author of *Who Do You Want Your Customers to Become?*, *The Innovator's Hypothesis: How Cheap Experiments Are Worth More Than Good Ideas* (MIT Press), and other books.